UNTANGLING TWINNING

NOTRE DAME STUDIES IN MEDICAL ETHICS AND BIOETHICS
O. Carter Snead, series editor

The purpose of the Notre Dame Studies in Medical Ethics and Bioethics series, sponsored by the de Nicola Center for Ethics and Culture, is to publish works that explore the ethical, cultural, and public questions arising from advances in biomedical technology, the practice of medicine, and the biosciences.

UNTANGLING TWINNING

What Science Tells Us about the
Nature of Human Embryos

MAUREEN L. CONDIC

University of Notre Dame Press
Notre Dame, Indiana

University of Notre Dame Press
Notre Dame, Indiana 46556
undpress.nd.edu

Library of Congress Cataloging-in-Publication Data

Names: Condic, Maureen, author.
Title: Untangling twinning : what science tells us about the nature of human embryos / Maureen L. Condic.
Other titles: Notre Dame studies in medical ethics and bioethics.
Description: Notre Dame, Indiana : University of Notre Dame Press, [2020] | Series: Notre Dame studies in medical ethics and bioethics | Includes bibliographical references and index.
Identifiers: LCCN 2019054557 (print) | LCCN 2019054558 (ebook) | ISBN 9780268107055 (hardback) | ISBN 9780268107086 (adobe pdf) | ISBN 9780268107079 (epub)
Subjects: MESH: Twins | Beginning of Human Life—ethics | Embryonic Development | Bioethical Issues | Chimerism—embryology
Classification: LCC RG133.5 (print) | LCC RG133.5 (ebook) | NLM WQ 235 | DDC 176—dc23
LC record available at https://lccn.loc.gov/2019054557
LC ebook record available at https://lccn.loc.gov/2019054558

This book is dedicated to Joseph Yost, who has supported me with critical scientific discussion, tireless devotion, and a never-failing Irish sense of humor.

CONTENTS

ILLUSTRATIONS

TABLES

ACKNOWLEDGMENTS

Portions of this work or the ideas expressed herein have been adapted from previously published texts, including but not limited to the following:

Condic, M. L. "Life: Defining the Beginning by the End." *First Things* 133 (May 2003): 50–54.

Condic, M. L., and S. B. Condic. "The Appropriate Limits of Science in the Formation of Public Policy." *Notre Dame Journal of Law, Ethics and Public Policy* 17, no. 1 (2003): 157–79.

Condic, M. L., and S. B. Condic. "Defining Organisms by Organization." *National Catholic Bioethics Quarterly* 5, no. 2 (2005): 331–53.

Condic, M. L. "When Does Human Life Begin? A Scientific Perspective." Westchester Institute White Paper (Westchester Institute for Ethics and the Human Person) 1, no. 1 (October 2008): 1–18.

Condic, M. L. "Alternative Sources of Pluripotent Stem Cells: Altered Nuclear Transfer." *Cell Proliferation* 41, suppl. 1 (December 2008): 7–19.

Condic, M. L. "Preimplantation Stages of Human Development: The Biological and Moral Status of Early Embryos." In *Is This Cell a Human Being? Exploring the Status of Embryos, Stem Cells and Human-Animal Hybrids*, edited by Antoine Suarez and Joachim Huarte, 25–43. New York: Springer, 2011.

Condic, M. L. "A Biological Definition of the Human Embryo." In *Persons, Moral Worth, and Embryos: A Critical Analysis of Pro-Choice Arguments*, edited by Stephen Napier, 211–35. New York: Springer, 2011.

Condic, M. L. "The Science and Politics of Cloning: What the News Was All About." *On Point*, Charlotte Lozier Institute, May 1, 2013.

https://s27589.pcdn.co/wp-content/uploads/2013/05/On-Point
-Science-and-Politics-of-Cloning-Condic-May-2013.pdf.

Condic, M. L. "Human Embryology: Science Politics versus Science Facts." *Quaestiones Disputatae* 5, no. 2 (2014): 47–60.

Condic, M. L., and K. Flannery, "A Contemporary Aristotelian Embryology." *Nova and Vetera* (English Edition) 12, no. 2 (2014): 495–508.

Condic, M. L. "When Does Human Life Begin? The Scientific Evidence and Terminology Revisited." *University of St. Thomas Journal of Law and Public Policy* 8, no. 1 (2014): 44–81.

Condic, M. L. "Totipotency: What It Is and What It Is Not." *Stem Cells and Development* 23, no. 8 (April 2014): 796–812.

Condic, M. L. "Determination of Death: A Scientific Perspective on Biological Integration." *Journal of Medicine and Philosophy* 41, no. 3 (June 2016): 257–78.

Condic, M. L. "The Role of Maternal-Effect Genes in Mammalian Development." *Stem Cell Reviews and Reports* 12, no. 3 (June 2016): 276–84.

Condic, M. L. "Embryos and Integration." In *Life and Learning XXVI: Proceedings of the Twenty-Sixth University Faculty for Life Conference*, edited by Joseph W. Koterski, 295–323. Bronx: Fordham University Press, 2017.

Condic, M. L. "Virtues beyond a Utilitarian Approach in Biomedical Research." In *Proceedings of the XXII PAV General Assembly*, 99–113. Rome: Libreria Editrice Vaticana, 2017.

Condic, M. L., and S. B. Condic. *Human Embryos, Human Beings: A Scientific and Philosophical Approach*. Washington, DC: Catholic University of America Press, 2018.

NOTE ON THE PRESENTATION
OF CITATIONS IN THE TEXT

Some endnotes expand upon points that are stated briefly in the main text or refer readers to other sections of the book that discuss the topic in greater detail. You can identify endnotes that contain additional discussion, clarification, or an internal reference by the brackets flanking their superscripted callout numbers in the text.

INTRODUCTION

Human Embryos and Human Individuals

For the vast majority of human history, prenatal development was a deep mystery that could not be penetrated by direct observation. While scientists, philosophers, and bioethicists have considered the origins of human life for a long time (for example, Aristotle discusses embryonic development extensively in *De generatione animalium*), the conclusions they have drawn were often based on very little evidence. Consequently, appealing to historical "experts" yields a plethora of opinions, many of which have very little to do with the scientific facts.[1]

In modern times, with the advent of chemical contraception, in vitro fertilization (IVF), and human embryo research, determining precisely when human life begins has become a matter of considerable importance. Each of these practices raises significant questions regarding the nature of the entity produced by sperm-egg fusion and society's obligation to that entity. The ability to manipulate the earliest stages of human life in the laboratory has brought into sharp focus a number of questions that are vital to our understanding of human beings and human rights, including the following: When does human life begin? Is a human embryo a human individual? What is the basis of human value?

There are no universally agreed-upon answers to these questions. Life is clearly a continuum, with living cells giving rise to new types of cells

1

and, ultimately, to mature individuals. This fact has led many to conclude that it is impossible to determine when human life begins and to question whether human embryos have greater value than human cells. Yet this view raises a serious ethical dilemma: while no one objects to the destruction of ordinary human cells for biomedical research, the use of human beings for such purposes is universally condemned. To resolve this dilemma, clear criteria must be established to determine when living human cells give rise to a new individual human being.

The phenomenon of identical (monozygotic) twinning presents a significant challenge to the view that human life and human personhood begin at conception.[2] The fact that a single embryo can split to generate two (or more) genetically identical embryos seems to defy the notion that prior to splitting, the embryo can itself be an individual human being. The fundamental philosophical challenge of twinning is an ontological one; if a one-cell embryo (or zygote) that would normally mature into a single individual can split early in development to give rise to two embryos, this calls the ontological status of the original zygote into question. A single cell cannot be simultaneously one individual and two individuals. Consequently, many have concluded that so long as the potential for identical twinning exists, no single human individual can exist.

The view that no human individual can exist so long as twinning is possible has led to a widespread denial of the individual humanity of early human embryos, particularly within the scientific community. This view of the embryo was initially promulgated in 1979 by biologist Clifford Grobstein, chairman of the biology department and the dean of the School of Medicine at the University of California, San Diego. In defense of the newly pioneered practice of IVF, Grobstein argued that the procedure did not produce an embryo, but only something that would eventually become an embryo—and he coined a new term to describe this entity: a "preembryo."[3] Interestingly, Grobstein himself does not use the term "preembryo" in a related article with nearly an identical title, published four years later in the prestigious scientific journal *Science*, instead referring to the product of IVF as an "embryo."[4] Yet despite this inconsistency and despite significant opposition to the term "preembryo,"[5] Grobstein and other biologists[6] remained strongly committed to the view that the early human embryo was not a human being and was, therefore, an appropriate subject of destructive scientific experimentation.

Nearly a decade later in 1988, Grobstein attempted to clarify precisely what distinguishes a "preembryo" from an embryo, indicating that his term was "intended to designate the period from fertilization to the first visible sign of the formation of the actual embryo, the so-called primitive streak."[7] Formation of the primitive streak takes place at approximately fourteen days of development and marks the point beyond which twinning is no longer possible. In characterizing the "preembryo," Grobstein states:

> I will begin by listing the fundamental biological characteristics of the stages here designated as preembryonic. First, the post-fertilization period involves a new genetic individual resulting from fusion of gametes from two human parents. The zygote and subsequent stages are thus indisputably alive, human, and genetically individual (unique). By virtue of their genetic composition, they are hereditarily related to others who are their kin.[8]

By Grobstein's own characterization, it is difficult to distinguish an embryo from a "preembryo," raising the question of whether there is a scientifically meaningful difference between the two. To address this possible objection, Grobstein goes on to define a number of features that he believes distinguish a "preembryo" from a human being, with the first and most significant being the possibility of monozygotic twinning. Because a "preembryo" can be experimentally split to generate more than one individual, Grobstein concludes:

> While fertilization establishes genetic individuality, it does not establish a second aspect of individuality, namely oneness or singleness. This aspect of individuality may be called developmental individuality because, without it, the preembryo would not develop into an integrated and single adult.[9]

It must be noted that developing into "an integrated and single adult" is precisely what the vast majority of embryos that do not undergo twinning actually do, so it is difficult to imagine what aspect of "oneness" they lack. Yet despite the confused nature of Grobstein's logic, his terminology, in combination with an influential book by Fr. Norman Ford that made a similar argument regarding twinning and individuality,[10] greatly impacted modern thinking and played an important role in the development of research policies regarding human embryos.

The view that an embryo is merely a human cell or cluster of human cells on the way to becoming a human being affords embryos little or no moral value. Many institutions and countries have adopted this view, establishing research policies that allow experimentation on human embryos prior to "individuation," or the point at which twinning is no longer possible. For example, in the United Kingdom, the 1984 report of the Warnock commission[11] held that the use of human embryos in research was permissible up until day fourteen of development, based in part on the assertion that cells remained "undifferentiated" prior to this point (an assertion that was clearly contradicted by substantial data, even at the time). The Warnock commission explicitly endorsed the term "preembryo" to describe the early stages of human development. And although the term "preembryo" has not been adopted by the scientific community,[12] this view has nonetheless persisted in many contexts—for example, recent textbooks on bioethics,[13] epigenetics,[14] and law.[15]

Based on the Warnock report, license was granted in the United Kingdom (and subsequently in the United States) to use public funding for research on human embryos prior to the fourteenth day of development.[16] Since that time, we have witnessed the destruction of human embryos to produce embryonic stem cells,[17] to clone human beings,[18] to manufacture human embryos with three biological parents,[19] to permanently alter the human genome[20] and to produce human-animal chimeras in which more than half of the brain is composed of human cells.[21] In addition to legitimate questions regarding the value of these experiments,[22] the sheer numbers of human embryos that have been created and destroyed for medical and research purposes is astonishing. In the United Kingdom alone, the Human Fertilization and Embryology Authority estimates that between 1991 and 2012, over 1.7 million "spare" embryos were discarded following fertility treatments.[23] The number of human embryos destroyed for research purposes is unknown.

To formulate sound policy regarding biomedical research involving human embryos, it is important to have a clear understanding of the scientific evidence relevant to both the beginning of human life and human twinning, as well as a sound view of the human individual as the subject of human rights. Here I will review the scientific evidence regarding early human development and human twinning, and then address the main philosophical problems raised by twinning in light of this evidence.

CHAPTER 1

WHEN DOES HUMAN LIFE BEGIN?

The Origin of New Cells

In considering the question of when the life of a new human being commences, we must first address the more fundamental question of when a new cell, distinct from sperm and egg, comes into existence. Human cells can be distinguished from each other by scientific criteria. Indeed, the entire field of biology is based on the ability of scientists to distinguish one cell type from another. For example, skin cells can be converted into pluripotent stem cells by manipulation of specific genes during cellular reprogramming,[1] but this is clearly a *conversion* of one cell type to another. No credible scientist would argue that skin cells are already pluripotent stem cells or are the equivalents of pluripotent stem cells. These are two distinct cell types with distinct properties. The fact that one cell type can give rise to a different cell type in no way alters the fact that a new cell type has been produced.

How do scientists determine when a new cell type has been produced, either in the laboratory or as a consequence of a natural biologic process? The scientific basis for distinguishing one cell type from another rests on two criteria: differences in molecular composition and differences in cell behavior.[2] Differences in molecular composition can arise due to an alteration in gene expression, a change in the subcellular localization of existing molecules, or a chemical modification of existing molecules. Alternatively, when cells exhibit new behavior, for example, going from a stationary to an actively migratory state, they can also be identified as distinct cell types. In many cases, changes in composition directly cause changes in behavior. Importantly, the criteria for when a new cell forms are based on scientific observations (not mere opinion or speculation): observations that are employed throughout the scientific enterprise and that can be independently verified.

Based on these criteria, the fusion of sperm and egg clearly produces a new cell type. Following the binding of sperm and egg to each other, the membranes of these two cells fuse, creating a single hybrid cell: the zygote or one-cell embryo.[3] Cell membrane fusion is a well-studied and very rapid event, occurring in less than a second.[4] Because the zygote arises from the fusion of two different cells, it contains all the components of both sperm and egg, and therefore the zygote has a unique molecular composition that is distinct from either gamete.

Subsequent to sperm-egg fusion, events rapidly occur in the zygote that do not normally occur in either sperm or egg.[5] Within seconds, the zygote initiates a molecular cascade that will, over the next thirty minutes, result in chemical modifications that prevent additional sperm from binding to the cell-surface. Thus, the zygote acts immediately and specifically to antagonize the function of the gametes from which it is derived. The "goal" of both sperm and egg is to find each other and fuse. Yet the first act of the zygote is to prevent any further binding of sperm to the cell surface. Clearly, the prior trajectories of sperm and egg have been abandoned, and a new trajectory—that of the zygote—has taken their place.

Based on this factual description of the events following sperm-egg fusion, we can confidently conclude that a new cell (the zygote), distinct from the gametes giving rise to it, comes into existence at the scientifically well-defined "moment" of sperm-egg fusion.

The conclusion that sperm-egg fusion creates a new cell has been questioned by Eugene Mills,[6] who argues that subsequent to sperm-egg fusion, only a modified gamete, or "fertilized egg" exists, with eggs and fertilized-eggs being ontologically "identical." Mills's position has been effectively rebutted by Calum Miller and Alexander Pruss,[7] who argue that (1) it is not sufficient to claim identity of the zygote and the oocyte based on the fact that the word "egg" is at times used to describe both entities, (2) the fact that at sperm-egg fusion an oocyte appears to persist when viewed with light microscopy is insufficient to reveal the many relevant changes that occur at a submicroscopic level, and finally, (3) there is no category other than an uninformative and nondiscriminating gross category such as "thing" or "cell" (a "coarse-grained chunk of matter"[8]) that can define an oocyte and a zygote as identical—categories that would also define as "identical" a living human and the corpse of that individual after death. Based on the argument of Miller and Pruss as well as on the criteria used by the scientific community to distinguish different cell types, the phrase "fertilized egg" is clearly not a valid scientific term, and the conclusion that sperm-egg fusion produces a new kind of living human cell is warranted. Yet the important question of what kind of cell is produced by this event remains to be addressed.

What Kind of Cell Is Produced by Sperm-Egg Fusion?

What is the nature of the new cell that comes into existence upon sperm-egg fusion? Most importantly, is the zygote a new human individual or, as Grobstein maintained, merely a human cell that will eventually give rise to a new human individual—that is, a "preembryo"? Both human cells and human organisms are alive, exhibiting characteristics that are uniquely observed in living entities, such as autonomy, integration, and reproduction.[9] Yet just as science distinguishes between different *types* of cells, it also clearly distinguishes between human cells that are *parts* of an integrated organism and *whole* human organisms.

An organism is defined as "(1) a complex structure of interdependent and subordinate elements whose relations and properties are largely determined by their function in the whole and (2) an individual constituted to carry on the activities of life by means of organs separate in function but mutually dependent: a living being."[10] Based on this definition, the interaction of parts in support of an integrated whole is the distinguishing feature of an organism. Because organisms are "living beings," another name for a human organism is a "human being."

Human organisms have a number of traits that distinguish them from human cells. First, humans, and all multicellular organisms, undergo *development*—that is, an orderly sequence of maturation that results in a characteristic adult form. In contrast, cells may proliferate to produce copies of themselves or may generate tumors that contain many different cell types, but they will not produce a characteristic sequence of events that robustly results in the generation of a species-specific mature state. *Development* is the defining characteristic of all embryos from the zygote stage onward.[11]

A second characteristic of organisms is their ability to *repair* injury to restore the health and function of the entity as a whole. Embryos are quite remarkable in their ability to recover from even catastrophic injury. For example, when IVF-produced embryos are subjected to preimplantation genetic diagnosis, one cell is removed at the eight-cell stage. Surprisingly, in many cases, the embryo will regenerate the missing parts and proceed with normal development.[12] In contrast, when portions of tumors are removed, the remaining cells will continue to proliferate, but will not specifically regenerate the damaged structures or restore the original state.

A third characteristic of organisms is *adaptation* to changing environmental circumstances, such that the health and overall function of the organism is preserved. A striking example of embryo adaptation is ectopic implantation. In most cases, implantation outside of the uterus is not compatible with life; embryos typically fail to establish an adequate placental circulation with the mother and subsequently die. There are cases, however, where healthy, term infants have been delivered after implanting in their mother's ovary,[13] abdomen,[14] or liver,[15] indicating that embryos can adapt to highly abnormal environments and continue to mature normally.

Finally, at all stages of the life span, organisms show *integrated function of parts* to promote the health of the organism as a whole. Within minutes of sperm-egg fusion, the zygote initiates a specific molecular cascade, using elements derived from both sperm and egg, to direct its subsequent maturation.[16] Substantial evidence indicates that by the four-cell stage (or earlier[17]), individual blastomeres have distinct patterns of gene expression,[18] different cellular functions,[19] and unique developmental capabilities.[20] This indicates that the integrated function of the embryo as a whole reflects collaboration of its parts to generate a normal developmental sequence. While all cells are capable of short-range cell-cell communication, only embryos establish a global pattern of interaction that benefits the entity as a whole. Indeed, within the first minutes and days of life, the embryo initiates dozens of distinct, globally integrated events that are critical for its survival and healthy maturation.[21]

Based on this scientific description of fertilization and subsequent embryonic development, sperm-egg fusion clearly generates a new human cell, the zygote, with composition and behavior that are distinct from that of either gamete. Moreover, the zygote is not merely a unique human cell, but a cell with all the properties of a fully complete (albeit immature) human organism; it is "an individual constituted to carry on the activities of life by means of organs separate in function but mutually dependent: a living being."

This understanding of the embryo provides the foundation for a broader view of what it means to be human. As human organisms, we respond to our environment in a manner that maintains the survival and health of our bodies as a whole. This integrated function and "wholeness" are intrinsic to our nature and are the basis for any ethical view of human beings, as distinct from human cells. The facts of human embryology outlined here allow us to establish a first unifying concept for consideration of the human embryo:

UNIFYING CONCEPT 1: Based on objective, scientific criteria, the life of an individual human being unambiguously begins at sperm-egg fusion.

CHAPTER 2

TOTIPOTENCY

What Is Totipotency?

The term "totipotency" is widely used in the scientific literature to describe the ability of a zygote to generate the mature body. Importantly, totipotency is a property of a *single cell*, not a group of cells; that is, a later-stage embryo, composed of many cells, retains the ability to generate the mature body, but this is a property of the embryo *as a whole*, not a property of an individual cell within the embryo. Consequently, the term "totipotent" applies exclusively to a one-cell embryo. Yet surprisingly, many scientific authors hold that totipotency is *not* a unique feature of zygotes and therefore conclude that because zygotes are not unique, they cannot be afforded a unique moral status.[1]

Whether cells that are not zygotes are also totipotent depends critically on the meaning of the term. "Totipotency" has two definitions: (1) "capable of developing into a complete organism" and (2) capable of "differentiating into any of [a complete organism's] cells or tissues."[2] Much of the confusion surrounding the use of the term "totipotent" centers on the important difference between these two definitions. A zygote is clearly "totipotent" in both senses, yet some authors characterize tumors[3] and stem cells[4] as "totipotent," based only on the second definition. In contrast, "pluripotency" refers to the ability of a cell to produce all of the cell types of the mature body, but not the cells of the placenta and embryonic membranes. Thus, embryonic stem cells from some species[5] are pluripotent, but not totipotent in either of the two senses given above.

To produce a mature organism, the zygote must both *generate* all the cells of the body and also *organize* them in a specific temporal and spatial sequence. Totipotency in this strict sense is experimentally demonstrated by isolating a cell in a supportive environment and seeing if it is able to produce a fertile, adult individual entirely on its own. For humans (and other mammals), this would involve transferring an isolated cell to the uterus and determining whether this results in pregnancy and live-birth.[6] Importantly, *any* totipotent cell, regardless of its origin, is *also* a one-cell embryo—that is, a cell that is capable of generating a globally integrated developmental sequence.[7] While stem cells, tumors, and embryos have many molecular features in common, only totipotent one-cell embryos (zygotes) are capable of *development*.

As discussed in detail elsewhere,[8] the fact that totipotency in the strict sense is the defining characteristic of a one-cell embryo has important ethical and social ramifications:

> Regardless how individuals or societies ultimately weigh the value of the embryo relative to the value of scientific research, it is important to appreciate that in all cases, the ethical consideration given to human embryos does not reflect the status they will achieve at some point in the future (i.e., what they will mature into). If this were the case, there could be no possible objection to embryo-destructive research since, by definition, adult status is never attained in such situations. Rather, ethical consideration is given to human embryos based on the status they already possess; i.e., their unique and fully operative ability to function as a human organism. Therefore, ethical controversy regarding totipotent human cells only concerns cells that are totipotent in the strict, organismal sense; i.e., a cell that is a human embryo.[9]

The conclusion that status as a human being does not depend on obtaining some future state, but rather on the nature of the human individual, is true of all stages of human life. For example, we grant special ethical status, including legal personhood, to infants not based on the fact that they will ultimately mature into adults, but rather because, *as infants*, they are already fully complete (albeit immature) human beings. This leads us to a second important unifying concept regarding human embryos:

> **UNIFYING CONCEPT 2:** The moral and ethical consideration due to a human embryo is based on the underlying nature of the embryo itself, rather than on the mature state it will ultimately achieve.

How Long Does Totipotency Persist?

The early cells of the embryo produced by cleavage of the zygote are known as "blastomeres" (figure 2.1). It is often asserted that totipotency is a feature of many early blastomeres, persisting even until the blastocyst stage.[10] However, there are very few empirical data to support this claim, in part because it is surprisingly difficult to determine how long totipotent cells actually persist during development. As noted above, totipotency is experimentally demonstrated by the ability of an isolated cell to produce a fertile, adult individual (table 2.1). Yet testing potency in this manner is challenging; isolated cells often are damaged or otherwise unhealthy and do not survive long enough for researchers to determine their potency. And such a negative result is not informative.

As an alternative approach, some claim to demonstrate totipotency of blastomeres by injecting them into embryos that have been manipulated such that the developmental capacity of the embryo is impaired,[11] but not entirely eliminated.[12] Yet this procedure does not test the developmental potency of the blastomeres themselves. Rather, it tests the developmental potency of a manipulated embryo in combination with normal blastomeres. Consequently, such assays are not a legitimate test for totipotency.[13]

Based on cell-isolation experiments conducted largely in mice, it appears that only the first two cells of the mammalian embryo remain totipotent—that is, able to generate a complete individual on their own.[14] A recent study in mice of over a thousand pairs of blastomeres isolated

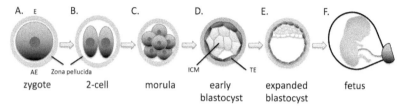

Figure 2.1. Preimplantation human development. (**A**) The one-cell embryo, or zygote, forms at sperm-egg fusion. The side of the embryo that will form the inner cell mass (ICM) is known as the embryonic pole (E), with the opposite side known as the abembryonic (AE) pole. Early development occurs within an acellular protein coat known as the zona pellucida (outer grey circle). (**B**) The zygote divides to form the two-cell embryo at approximately twenty-four to thirty hours. Individual cells are known as blastomeres. (**C**) The morula, or eight-cell embryo, forms on day two-three. (**D**) On day three-four, cells begin to adhere more tightly to each other, forming the early blastocyst, which fills with fluid and gives rise to the first two distinct cell types at the thirty-two-cell stage. The outer cells (dark gray) are trophectoderm (TE). The inner cells (light gray) are inner cell mass (ICM). (**E**) Cell division continues, and implantation occurs at the expanded blastocyst stage (day five-six), when the ICM has divided into epiblast (light gray) and hypoblast (white). (**F**) The postnatal body of the fetus (not shown to scale) is formed largely from ICM, while the placenta and membranes form largely from TE. An earlier version of this diagram appeared in M. L. Condic and S. B. Condic, *Human Embryos, Human Beings: A Scientific and Philosophical Approach* (Washington, DC: Catholic University of America Press, 2018).

at the two-cell stage demonstrates that in the majority of cases, only one cell of a two-cell embryo retains the ability to undergo development on its own.[15] A significant caveat to these studies is that failure of an isolated cell to undergo normal development is a negative finding with limited explanatory power.[16] It has been suggested that limited potency of later blastomeres may be due to their relatively small size, compared to the zygote.[17] Potency also appears to vary quite a bit between species. For example, live-born cattle are routinely generated from cells separated at the four-cell stage,[18] while live-born mice produced from cells isolated at the two-cell stage are rare.[19] There has also been a single (unreplicated) report of totipotency in pigs persisting until the eight-cell stage,[20] but there is no evidence for totipotent cells persisting in any mammalian species beyond this point. These data suggest, but do not prove, that

Table 2.1. Tests for progressive developmental restriction

Test	What is tested	Interpretation of test
Cell isolation and subsequent development	Potency	The range of cells produced by an isolated cell defines the developmental "power" of that cell, internal to itself. Failure to produce specific cell types suggests, but does not prove, a potency restriction.
Lineage analysis within an intact embryo	Fate	What a cell naturally produces reflects its developmental potency. If fate is restricted, this suggests, but does not prove, potency restriction. If fate is random, this proves potency is not restricted.
Differences in gene expression	Specification	Differences in gene expression prove that cells have entered into distinct developmental pathways. Potency may not be restricted; i.e., a specified cell may still be capable of shifting to a different developmental path.
Cell transplantation or reaggregation	Commitment	The response of a cell to a new environment determines whether potency has been restricted, such that the cell is now unable to shift to a new developmental path.

in humans, totipotency is unlikely to be preserved beyond the two-cell stage, and almost certainly does not persist beyond the four-cell stage.

This does not mean that the embryo at the two- or four-cell stage is composed of two or four *zygotes*. If this were the case, a single embryo would never develop, but would only continue dividing to generate an infinite collection of totipotent zygotes.[21] Rather, the evidence indicates that early blastomeres normally function as parts of the embryo, producing a limited set of progeny, yet they can nonetheless *become* totipotent zygotes if separated from the embryo and cultured on their own.

As an alternative to directly testing the potency of cells by isolation, researchers also examine cell "fate," or the lineage produced by a specific cell in an intact embryo (table 2.1, Fate). This is done by injecting a stable marker (e.g., a dye) into a single cell and determining the range of labeled cell types produced at a later developmental time (i.e., the cell lineage). There are two possible outcomes for this experiment. First, the lineage may be *determined*, or identical in all members of the species. This finding is not informative, since a consistent lineage can reflect either a limitation in what the cell is able to produce (i.e., its potency) or merely an invariant set of developmental events that limit the nature of the cell's progeny. In contrast, if the lineage is random, this clearly demonstrates potency has not been restricted.

In considering mammalian cell lineages, here too we find there is not a simple answer to the question of when potency restriction occurs. Mouse embryos are by far the most well-studied mammalian species. Beginning in the first few years of this century, multiple groups demonstrated a consistent relationship between the embryonic-abembryonic axis of the mouse blastocyst and specific positions within the oocyte (figure 2.1A),[22] suggesting that some elements of body-axis may be biased (or even predetermined) by factors that are differentially distributed in mouse eggs. Similarly, careful analysis of mouse development indicates that in 80 percent of embryos, the lineages produced by the first two cells are distinct, but overlapping; that is, each of the first two cells is biased to produce a different subset of cells at more mature stages.[23] This finding also suggests, but does not prove, that potency is normally restricted by the two-cell stage.

A third, and again indirect, method of examining cell potency is to look for differences in gene expression among cells of the early embryo (table 2.1, Specification). Distinct patterns of gene utilization do not prove that developmental capabilities have been restricted, but they do strongly indicate that the cells have entered into distinct developmental pathways. Work from several groups has described regularly occurring molecular and/or functional differences between mammalian blasto-meres as early as the four-cell stage.[24] These differences become more pronounced by the eight-cell stage,[25] indicating that cleavage-stage mam-malian blastomeres undergo specification and that subsequent mam-malian development is to some extent mosaic. Consistent differences in gene expression occurring even earlier, at the two-cell stage, have re-cently been demonstrated by meta-analysis.[26] Importantly, these findings strongly suggest that since cells rapidly become "specialized," totipotency is unlikely to persist beyond the two-cell stage.

Yet, even in cases where lineage is restricted and cells exhibit differ-ences in gene expression, it is still not clear that potency has also been re-stricted—that is, that cells are *committed* to a specific developmental path. To test commitment, researchers transplant cells of the embryo from one location to another or dissociate and reaggregate entire embryos to see whether cells will shift to a new developmental path when placed under the influence of new developmental signals (table 2.1, Commitment).

Like cell-isolation experiments, cell transplantation/reaggregation to test commitment is challenging, because damaged cells may yield an un-informative negative result. However, several authors have asserted that blastomeres isolated from morula-stage embryos (eight to sixteen cells) are not yet committed because when they are reaggregated with blasto-meres that have been manipulated to be incapable of development on their own, either live-born animals or what appear to be normal blasto-cysts can be produced.[27] Yet this approach raises concerns similar to those expressed above regarding injection of blastomeres into manipulated em-bryos, because it does not test the potency of individual blastomeres or even a homogeneous group of blastomeres, but rather tests the potency of a diverse group of both manipulated and unmanipulated blastomeres.

Simple aggregation experiments are somewhat more informative. When "outer" cells (presumptive TE; figure 2.1) of early human blasto-cysts (sixteen-cell stage) are aggregated, they form blastocyst-like struc-tures that initiate expression of ICM-associated genes,[28] suggesting the potency of these cells has not been restricted to the TE lineage. In con-trast, when inner or outer cells taken from embryos at the thirty-two-cell stage are aggregated, live-born mice are not obtained,[29] suggesting that blastomere potency is restricted (i.e., cells are committed to either the TE or ICM lineage and cannot regenerate a complete embryo). Finally, when freshly isolated ICM[30] or TE[31] cells from blastocyst-stage embryos (thirty-two+ cells) are transferred to the uterus or other supportive locations, they do not undergo development, again suggesting that cells are committed to specific fates and cannot reconstitute a whole embryo on their own. Although there are significant caveats to the interpretation of all of these experiments,[32] collectively they suggest that cells of the mouse embryo become *committed* to specific developmental pathways between the sixteen- and the thirty-two-cell stage (in humans, roughly the third day of development), and therefore potency has been restricted by this point.

As a whole, this body of evidence clearly illustrates that development is an *ongoing, integrated process*; that is, the zygote gives rise to cells with distinct properties by the four-cell stage, and these cells become *committed* to specific paths by the thirty-two-cell stage. Importantly, the ability of early (eight-to-sixteen-cell stage) blastomeres to switch between the ICM and TE lineages demonstrates they remain "totipotent" in the less restrictive sense (i.e., able to generate all the cell types of the body) but *does not* demonstrate totipotency in the more restrictive, embryo-defining sense (i.e., able to both generate all cell types and organize them in a coherent, developmental sequence). The failure of isolated blastomeres to undergo development independently strongly suggests that totipotency in the strict sense does not persist beyond the two-cell stage.

Oocyte Cytoplasm Is a Critical Component of Totipotency

The fact that strict totipotency is lost so swiftly during development reflects the fact that cytoplasmic components of the oocyte[33] are critical for totipotency. Oocytes are unique cells with a characteristic pattern of gene activation[34] that is distinct from the pattern observed after fertilization or during early development.[35] Recent work has documented multiple oocyte-derived components that are essential for mammalian development. In extreme cases, the loss of a single maternal factor results in death of the embryo while having no impact on the health of the mother. Such developmentally required factors are known as maternal effect genes (MEGs).

MEGs are a fascinating class of genes because they are specifically re-quired for development, while being nonessential for any other biologic function. Over sixty oocyte-specific MEGs that regulate a wide range of developmental processes have been identified in mammals.[36] These data indicate that factors uniquely associated with oocytes are essential for to-tipotency and suggest that blastomeres inheriting only a subset of these components cannot be totipotent.

These five lines of evidence (cell isolation, lineage analysis, differ-ences in gene expression, cell transplantation, and analysis of MEGs) all strongly suggest that totipotency in the strict sense, or the ability to inde-pendently initiate a developmental sequence that results in the produc-tion of a mature individual, is a distinctive feature of the one-cell embryo or zygote and is unlikely to persist beyond the two-cell stage.

The nature of totipotency, including how long it is likely to persist in human development, allows us to define a third important unifying con-cept for consideration of human embryos:

> UNIFYING CONCEPT 3: Totipotency is a property of a single cell, and a mature human oocyte is a critical component of totipotency. Therefore, only a cell directly derived from an oocyte (i.e., the zygote) or the imme-diate progeny of the zygote (if separated from the embryo as a whole) can be a totipotent human embryo.

CHAPTER 3

WHAT IS AN EMBRYO?

A clear understanding of the scientific evidence has allowed us to conclude that the life of a human individual begins at the instant of sperm-egg fusion. Yet this analysis leaves unaddressed the question of how an embryo is different from a nonembryo. At first inspection, the questions "When does life begin?" and "What is an embryo?" seem to be largely overlapping; that is, the fact that life begins at sperm-egg fusion would appear to define the product of sperm-egg fusion as a human organism. Indeed, when development proceeds normally, sperm-egg fusion clearly marks the beginning of a new human life.

Yet how are we to view cases in which development *does not* proceed normally or arrests entirely? If the cell produced by sperm-egg fusion does not mature into a healthy infant, how are we to view that cell? Is it a "defective embryo," or is it not an embryo at all? Moreover, how are we to view the plethora of cellular entities generated in the laboratory or in nature that have some similarity to embryos[1] but do not undergo normal human development, and therefore have unclear ontological status (table 3.1)?[2] Finally, the question of how we can reliably identify an embryo is distinct from the question of what an embryo actually is. When the cell produced by sperm-egg fusion arrests and/or does not proceed normally through development, it is impossible to determine the ontological

Table 3.1. Status of entities sharing some features in common with embryos

Natural or laboratory-generated human entities	Human nuclear genome	Arise(s) by fertilization	Can produce TE and ICM	Produce(s) a developmental sequence*	Human embryo
Zygote	+	+	+	+	Yes
Cloned embryo	+	-	+	+	Yes
Human-human chimera	+	-	+	+	Yes
Monozygotic twins	+	-	+	+	Yes
Cybrid embryo	+	+	+	+	Yes
Triploid or Tetraploid embryo	+	+	+	+	Yes; defective
Partial hydatidiform mole	+	+	+	+	Yes; defective
Complete hydatidiform mole	+	+	-	-	No
Parthenote	+	-	?	?	Insufficient data
Activated oocyte cytoplast	-	-	-	-	No
Embryonic stem cells (ESCs)	+	-	+	-	No
Induced pluripotent stem cells	+	-	+	-	No
Embryonal carcinoma	+	-	+	-	No
Teratoma/plenipotent tumor	+	-	+	-	No
Tissue explant	+	-	+	-	No
Adult stem cells	+	-	+	-	No
*Entities made in animals***					*Animal embryo*
Human-animal chimera‡	Minority of human cells	-	+	+	Ambiguous
Tetraploid complementation	n/a	-	+	+	Yes
Embryo splitting	n/a	-	+	+	Yes
Fused ESC-produced gametes	n/a	+	+	+	Yes
ANT product	n/a	-	-	-	No

*If there are *any* reported cases of this kind of entity forming TE/ICM in an ordered developmental sequence, it is listed as a positive.

**Entities made thus far only in animal models. Comparable human entities are likely to be similar.

‡Human/animal chimeras containing a minority of human cells develop along an animal trajectory. Whether shifting the ratio to a majority of human cells would alter this result is unknown.

status of that cell without clear criteria for identifying what is an embryo and what is not. Here I discuss two common views of how an embryo can be identified and the problems these views encounter, ending with proposed criteria that resolve these problems.[3]

Identifying a Zygote Based on Cell-Type of Origin

The most intuitive way of distinguishing embryos from nonembryos is based on how one-cell embryos are generated. Zygotes naturally arise as a consequence of sperm-egg fusion, so for many it seems reasonable to simply define a human zygote as an entity arising from such a fusion event. In light of this view, some have expressed concern that scientific experiments involving human egg cells must be prohibited, to avoid inadvertent production of embryos.[4]

While identifying a zygote as the product of a natural process (sperm-egg fusion) seems straightforward, this criterion is inadequate. Without clear molecular and (more importantly) *functional* criteria for what is and what is not a gamete, *any* organized biologic material derived from the gonads "could" be a sperm or an egg, if it exhibits *any* feature of a normal gamete. This kind of superficial similarity would include many entities that are not even living cells.

For example, if an egg cell is manipulated to remove the nucleus, the surviving cellular parts are known as a cytoplast. A cytoplast does not contain the necessary structures to support the full spectrum of life-functions and therefore it cannot be considered a living cell.[5] During natural maturation of human eggs, errors that result in abnormal numbers of chromosomes (including loss of the entire genome) are common.[6] Yet if a cytoplast entirely lacking a human genome is electrically or chemically stimulated (table 3.1, Activated oocyte cytoplast), it can undergo three rounds of division to generate a structure similar to a morula-stage embryo.[7] Thus a cytoplast that is naturally (albeit defectively) produced by the ovary can exhibit many characteristics of a developing embryo without even being a living human cell.

Identifying an embryo based solely on the tissue of origin (i.e., a cell produced by the fusion of cells derived from the ovary and testis) either pushes the issue back one step to the question "What is a gamete?" or creates a circular argument: an embryo is an entity produced by gametes, and gametes are those cells capable of producing an embryo.

Identifying a Zygote by Molecular Composition

A second common way of identifying a one-cell embryo is based on molecular composition. The clearest example of this approach is "systems biology."[8] Systems biology seeks to describe a living cell, including a zygote, based on a detailed analysis of all the molecules the cell contains and how they interact as a system. This would, of course, be an enormous undertaking. Yet even if such an analysis were possible, it would be fundamentally inadequate to robustly distinguish an embryo from a nonembryo.

The most significant problem with a systems biology approach is this: On what basis can we reliably determine which specific entities are to be included in a systems biology analysis? If, for example, we use cells arising from fusion of sperm and egg for our analysis of "embryos," then this selection *itself* constitutes an unexamined method of identifying an embryo that is not based on systems biology. Moreover, basing embryo selection on an intuitive, but inadequate criterion such as "the product of sperm-egg fusion" fundamentally compromises the accuracy of the analysis. For example, if an enucleated oocyte cytoplast fuses to a sperm, the resulting cell will have only half the characteristically human amount of DNA, and over time, it will either die or produce a disorganized tumor known as a complete hydatidiform mole (table 3.1). Clearly, the molecules, structures, and interactions observed in such a grossly abnormal entity cannot be legitimately considered characteristics of a human embryo.

The central problem with using systems biology to identify a zygote is that to determine the molecular composition of an embryo one must start with *embryos*, which requires independent criteria for what is and what is not an embryo. Thus, while a systems biology analysis may provide a wealth of interesting and useful *descriptive* information regarding the molecular composition of embryos, it is incapable of providing a *definition* of an embryo or a robust means to identify one.

Identifying an Embryo Based on Development

Human zygotes are distinct from all other human cells precisely because they are totipotent—that is, capable of development.[9] Therefore, an adequate definition of an embryo must also include a precise definition of the cellular and molecular events that constitute "development." It is important to note, however, that the inherent nature of an entity as either an embryo or a nonembryo is established at its initiation and does not depend on the gradual emergence of characteristic human traits. That is, development does not *produce* an embryo; rather, an embryo is an entity that is inherently capable of development.

Substantial evidence indicates that within seconds of sperm-egg fusion, a human zygote initiates a molecular cascade that regulates its subsequent maturation. Yet *all* cell types are capable of complex molecular cascades. Moreover, many of the molecular processes observed in stem cells and tumor cells are similar (or even *identical*) to processes occurring in an embryo. Despite such similarities, a tumor that generates a chaotic cellular mass containing a variety of different cell types is not undergoing development and is therefore not an embryo. As noted above, even an enucleated cytoplast, which is arguably not a *living cell*, can undergo three rounds of division, replicating many (but not all) features of normal embryogenesis. While the events occurring in the first hours and days after sperm-egg fusion are clearly *necessary* for embryonic development to proceed, are they *sufficient* for an entity to be considered an embryo?

To classify naturally occurring entities as well as those produced in the laboratory (table 3.1) based on the capacity to develop to a mature state of what it already is, I have proposed the following criteria for recognizing a developing one-cell human embryo:[10]

A human zygote is a discrete, living, biological entity with a human nuclear genome that
(1) initiates a globally organized and integrated (i.e., "organismal") developmental process having the potential to proceed up to or beyond the stage at which cells committed to trophectoderm and inner cell mass are present and

(2) has arisen from either

   (i)   the fusion of the plasma membranes of a human oocyte and a
         human sperm or

   (ii) any other event or procedure that initiates such an organized
         developmental process

   and

(3) has not yet proceeded through eight weeks of development since the
     initiation of such an organized developmental process.

In agreement with the large body of scientific evidence noted above, these criteria propose that a human embryo is initiated at a specific time ("a moment of conception"). In natural fertilization, this point would be the fusion of sperm and egg plasma membranes, yet any process that initiates development (such as cloning or embryo splitting) would also constitute a single event that marks the beginning of a new human life.

Two important aspects of these criteria are (1) entities formed by the fusion of gametes that do not initiate a developmental sequence are not human embryos and (2) formation of trophectoderm (TE) and inner cell mass (ICM; see figure 2.1) in an appropriate temporal and spatial sequence is both *necessary* and *sufficient* evidence for us to conclude that the entity is undergoing development—that is, to recognize that the entity is an embryo.

Appealing to the *behavior* of an entity as a way of understanding its nature reflects a long-standing intellectual tradition. Aristotle proposed that to know the nature of something, we must look at its "potentialities"—that is, its powers or capacities. Potentialities are revealed by acts—by what the entity does (Aristotle, *Treatise on the soul* 2.4.415a14–22). In this case, the entity formed at sperm-egg fusion reveals itself to be an embryo by undertaking the acts of development.

The Critical Role of Gametes in Establishing Strict Totipotency

The proposed criteria for recognizing an embryo presume a specific starting point, one that "initiates a globally organized and integrated (i.e., 'organismal') developmental process having the potential to proceed up to or beyond the stage at which cells committed to trophectoderm and inner cell mass are present."

In normal fertilization, this process is initiated by fusion of sperm and egg, unique cells that are naturally predisposed to forming a totipotent zygote. Gametes have a number of specific features that are necessary to produce an embryo upon fusion. Both sperm and egg contribute haploid genomes that are properly configured to participate in development (i.e., appropriately "imprinted"[11]). Moreover, as noted above, the oocyte contributes a large number of factors required for embryogenesis,[12] including factors that will appropriately modify the DNA derived from sperm and egg to establish an epigenetic state that is capable of development. These three gamete-dependent components (appropriately imprinted DNA, essential oocyte-derived factors, and an appropriate epigenetic state of the DNA) constitute the "program" for human development—the biological elements that distinguish a zygote from all other human cells.[13]

Because germ cells are naturally predisposed to make an organism upon fusion, entities derived from germ cells must be viewed with considerable caution, particularly in cases where there are insufficient data to resolve the nature of the entity (e.g., table 3.1, human parthenotes).[14]

In contrast, entities that are not derived from gametes and do not show globally coordinated behavior can be confidently determined to be nonembryos, because they are known to lack the gamete-derived elements that are required for totipotency. In some cases, tumors contain cells that are similar to TE and ICM or show what appears to be organismal behavior—for example, formation of a blastocyst-like structure[15] or expression of molecules in an embryo-like pattern.[16] Yet such a superficial resemblance to a developmental process is not evidence an embryo exists. As noted elsewhere,[17]

A number of studies have shown that ESC [embryonic stem cell] aggregates treated with specific signaling molecules exhibit some of the molecular cascades and cell behaviors observed during normal gastrulation. While these studies do not equate stem cells with embryos, several authors conclude that stem cell aggregates are surprisingly "embryo like." For example, one article states that aggregates "resemble normal embryonic development much closer than previously thought," exhibiting "an unexpected degree of self-organization." Yet stem cell aggregates do not resemble *organisms* so much as they resemble *tumors*, which can also show a surprising degree of self-organization, producing well-formed teeth and even, in one case, a re-markably normal eye.

Similarly, ESC aggregates and tumors occasionally generate cystic structures that have some visual similarity to a blastocyst-stage embryo, which has led some authors to ask whether ESCs and tumors might also be embryos. Yet many cell types with clearly restricted potency will form such cystic structures, including liver, heart, and cartilage; neurons; fibro-blasts; kidney; and umbilical cells, indicating that the mere formation of such structures is not sufficient evidence for totipotency.

In considering the ontological status of entities that are not derived from gametes, we must make a critical distinction between cells that ex-hibit *some* of the properties of cells within an embryo and cells that have *the same potency* as embryonic cells. If scientists were able to produce cell lines that stably exhibited the *full range* of natural properties observed in TE and ICM, then combining these cells would generate an embryo ca-pable of proceeding through development—just as viable embryos can be reconstituted after dissociation and reaggregation of blastomeres.[18] How-ever, given the critical contributions of gametes (particularly the oocyte) to totipotency,[19] it is extremely unlikely (if not impossible) for such cells to be produced without using gametes as the starting material.[20]

UNIFYING CONCEPT 4: Human gametes are highly specialized cells that are predisposed to form a human being upon fusion, yet not all cells produced by the ovaries and testes are functional gametes.

Integrating the Criteria for Recognizing a Zygote with the View of Humans as "Rational Animals"

The ability to reason is a distinguishing feature of human beings, and therefore the capacity for reason forms the basis of the classical defini-tion of human beings as "rational animals."[21] In light of the biological criteria for recognizing a zygote proposed here, a corresponding essential philosophical definition of a one-cell human embryo would be "a developing human," where the minimal criterion for "developing" is the production of TE and ICM as part of a globally integrated, organismal sequence. Thus, production of TE and ICM as part of a globally integrated, organismal sequence is necessary both to identify a cell as an embryo and for that cell to actually *be* an embryo.

A human organism at an embryonic stage has the active and proximate potency[22] to manifest rationality; that is, it actively self-develops into a rational animal and hence *is* one, even if specific defects ultimately prevent rationality from ever becoming operative. Biologically, having an "active and proximate potency to manifest rationality" requires that an entity possess the necessary molecules and structures to enable a self-directed trajectory of development that is ordered toward the production of the mature human body. Importantly, what is needed for rationality is *already present* from the zygote stage onward, even if the current state of development does not support the actual *operation* of reason. In an appropriately supportive environment, the totipotent zygote will, of its own inherent power, develop the structures necessary to manifest rationality.

Yet why is the ability to produce the first two committed cell types (TE and ICM) sufficient evidence that an entity possesses such an active, proximate potency to manifest rationality? Diverse operations are required for rational thought, and these diverse operations require a diversity of cell types.[23] Consequently, production of the first two cell types is the *minimal* evidence for possessing the active, proximate potency to produce the structures necessary to manifest rationality—that is, to undergo human embryonic development. Conversely, an entity that entirely lacks the ability to produce a diversity of cell types lacks an essential property of a human organism. Even in cases where a developing embryo fails to produce the structures necessary to support rationality (for example, the case of an anencephalic child who does not develop a normal brain), the potency to develop such structures is an inherent property of the entity, although it may be impeded by illness, injury, or a defect in the material structure of the embryo (see discussion below under "Why Not Later? Is the Generation of TE and ICM Sufficient for Recognizing an Embryo?"). As discussed elsewhere in more detail,[24] this requirement can be formally expressed as follows:

1. No entity entirely incapable of producing a body suitable for supporting rational thought is human.
2. Every entity entirely incapable of producing multiple cell types is incapable of producing a body suitable for supporting rational thought.
∴ No entity entirely incapable of producing multiple cell types is human.[25]

> UNIFYING CONCEPT 5: The capacity for development is the defining (i.e., essential) feature of a zygote, and formation of the first two committed cell types in a temporally and spatially integrated manner is the minimum criterion for identifying that this capacity exists.

Why Not Earlier? Is the Ability to Undergo Cell Specification Sufficient for Recognizing a Zygote?

If production of TE and ICM provides conclusive evidence that an embryo exists, how are we to view the stages of development prior to the formation of the first two cell types (figure 2.1)? Ample evidence indicates that at sperm-egg fusion a new cell is generated and that, in the course of normal human development, this cell *immediately* initiates an integrated, developmental trajectory.[26] Yet molecular differences between presumptive TE and ICM are not detected until at least the four-cell stage, and stable cell commitment occurs even later, between the sixteen- and thirty-two-cell stage (approximately four days after sperm-egg fusion). Clearly, many events characteristic of embryonic development occur prior to formation of TE and ICM. Why is the production of committed TE and ICM cells sufficient to identify an entity as a human organism, whereas these earlier events are not?

To address this question, it is important to appreciate that distinct biologic processes can nonetheless share common elements. As noted above (table 3.1), complete hydatidiform moles (CHMs) are formed from fusion of sperm and egg. CHMs share a large number of features in common with actual embryos, yet are clearly tumors, not organisms.

When an entity produced by gamete fusion undergoes *some* of the processes characteristic of early development, yet fails to replicate others, it is important to carefully consider the nature of such "failures" and to determine whether or not the entity exhibits a pattern of globally integrated (i.e., organismal) behavior.

Normally developing embryos produce new cell types in an orderly manner, with the first step in this process being "specification" (table 2.1). Cell specification indicates a cell has received information that directs it into a unique developmental path. Specification is detected by changes in cell composition (i.e., gene expression), and during the formation of TE and ICM, the earliest changes in gene expression occur at the four-cell stage. However, specified cells are not "committed" and are able to switch from one developmental path to another (table 2.1); that is, the potency of a specified cell is not yet restricted.

While cell specification is a *necessary* feature of an embryo, specification alone is not sufficient to confidently identify an entity as an embryo for at least two reasons. First, the sequence of molecular events associated with specification does not necessarily reflect a *developmental* sequence. Indeed, changes of the type observed during cell specification routinely occur in cell cultures, in tumors, and in tissues of the mature human body. A molecular change that merely results in cell specification but not differentiation/commitment can be either a developmental process or merely a cellular process.

Second, development requires that distinct cell types be produced and specification does not involve a restriction in cell potency; specified cells remain developmentally equivalent. Groups of functionally equivalent cells interacting with each other *can* reflect a developmental process, but does not *necessarily* indicate an embryo is present. It is only following the production of an appropriate pattern of cell commitment that a *developmental* process can be distinguished from a mere *cellular* process. Therefore, cell specification is necessary for committed cell types to arise and for an embryo to be present, but in and of itself, it is not *sufficient* to indicate that a developmental sequence is under way.

Central to this argument is the view that the level of organization an entity is capable of achieving allows us to determine the kind of entity it is. Isolated cells and unicellular organisms show *cellular* organization;

that is, the molecules and structures within the cell are ordered to maintain the life and health of the cell itself. Individual cells can, through processes similar to those seen in an embryo, produce tumors containing multiple cell types, but the cellular components of the tumor have no relationship to each other and do not function as an integrated whole. Therefore, a tumor is an *aggregate* of entities that show only cellular organization. In contrast, naturally formed tissues and organs show *tissue* organization; that is, two or more different cell types are ordered relative to each other to maintain the structure and function of the tissue. In some cases, individual cells can spontaneously organize as tissues.[27] The highest level of organization is *organismal,* where all the cells, tissues, and organs interact in a manner that is ordered to maintain the life and health of the entity as a whole.[28]

An important aspect of this line of reasoning is that the outcome of a biologic process is critical to interpreting the *nature* of that process. *Even if an initial sequence of molecular events is identical in two entities, when the outcome is different, the entities must also have different molecular composition.*[29] Importantly, entities with different outcomes do not become distinct at the point at which the first difference is observed. Rather, they have different natures from the beginning. Distinguishing between similar entities prior to observing differences in outcome can be challenging, but clearly such entities are not identical from their inception. Intrinsic differences in the molecular composition of the two entities *produce* the difference in outcome.

For example, a complete hydatidiform mole (CHM) is formed by fertilization of an egg that has abnormally lost its nucleus (a cytoplast), and therefore a CHM contains only paternally derived chromosomes. Current technology cannot reliably distinguish between a CHM and a zygote in a nondestructive way, but they are clearly distinct from the beginning. CHMs initially undergo many of the processes that are seen in normal embryos, but do not produce TE and ICM. Rather, CHMs produce tumors with only TE-like properties. Thus, despite being formed by fusion of a normal sperm and an "egg,"[30] and despite initiating complex molecular processes that are similar (and in some cases, *identical*) to a bona fide zygote, CHMs do not start out as embryos and later transform into tumors. They are *intrinsically* tumors ab initio.

A simple example helps illustrate this point. The song "Twinkle, Twinkle, Little Star" and the "Alphabet Song" are identical until the fourth measure (figure 3.1). If a recording of one of these two songs was stored as a computer file and then played for a naïve listener, there would be no question regarding when the song commenced; that is, the sounding of the first note would clearly mark the "instant" that the song began, and this note would exclude a large number of musical compositions with different starting points. Yet it would be *impossible* to accurately determine the identity of the song until the distinguishing notes of the fourth measure were played. Importantly, our inability to correctly *identify* the song is an epistemological, not an ontological, problem.[31] Only *one set of notes* is recorded in the file, despite our inability to discern the nature of the song until the fourth measure. Moreover, if we examined the storage device with sufficient resolution prior to playing the recording (using scanning tunneling microscopy, for example), it would be possible to determine the identity of the song from the beginning, because the information intrinsic to the computer file would differ between the two distinct songs.

Importantly, a recording of "Twinkle, Twinkle" is not somehow "thwarted" in its attempt to play the "Alphabet Song" by a deficiency of notes in the fourth measure. Nor is the recording "transformed" from the "Alphabet Song" into "Twinkle, Twinkle" by this deficiency. Nor is it the case that "neither song was initially playing" or that "both songs were initially playing" until the notes of the fourth measure sounded. *From the beginning*, the recording contains information specific to *only one song*, and that song was played from the first note onward.

Similarly, ab initio, a cell produced by sperm-egg fusion contains specific "information" (e.g., DNA, regulatory molecules, a particular epigenetic state, and a particular pattern of gene imprinting) that reflects the kind of cell it is. What occurs over time is simply the *manifestation* or "playing out" of the underlying information present in the cell, as specified by the cell's intrinsic nature.[32] Thus, even if the molecular events occurring in a CHM are presumed to be identical to those of a developing embryo,[33] a CHM does not start out as an embryo, only to be "thwarted" on the third day by a lack of maternally derived DNA. Rather, *from the beginning*, a CHM manifests its own inherent

Figure 3.1. Differences that can only be detected over time. What something is cannot necessarily be determined by the initial sequence of actions it undertakes. Recordings of "Twinkle, Twinkle, Little Star" and the "Alphabet Song" begin with the same notes and only diverge in the fourth measure, yet the files encoding these two songs are distinct (albeit very similar) from the beginning. An earlier version of this diagram appeared in M. L. Condic, "A Biological Definition of the Human Embryo," in *Persons, Moral Worth, and Embryos: A Critical Analysis of Pro-Choice Arguments*, edited by Stephen Napier (New York: Springer, 2011), 211–35.

properties—the properties of a tumor. It is simply the case that, like the two songs discussed above, CHMs and zygotes share a similar molecular composition and, therefore, a similar initial sequence of molecular events. Yet the disordered growth of a CHM is caused by its *own intrinsic nature*, a nature that does not include the capacity to function as an organism and to generate an integrated developmental sequence.

As argued above, the end result of a sequence of molecular events is not incidental to the interpretation of that sequence. Many of the early molecular events occurring in an embryo are consistent with *either* an embryonic process or a mere cellular process. Thus, while the initial events of development clearly define the instant at which both a zygote and a CHM are formed and clearly distinguish both of these entities from many other human cells, they do not distinguish them from each other. Formation of TE and ICM enables us to determine whether an embryonic or a cellular process is underway. Importantly, an embryo is not *made to be an embryo* by production of the first two cell types; both a CHM and a zygote have specific properties and potentialities ab initio. Rather, formation of

TE and ICM is simply the earliest point at which we can *discern* whether a developmental sequence has been initiated and (therefore) whether the entity under consideration is in fact a human organism.

> **UNIFYING CONCEPT 6:** If the necessary structures (molecules, genes, etc.) required for production of the first two committed cell types do not exist in an entity from the beginning, the entity is intrinsically incapable of undergoing development as an organism and is therefore not a human being.

Why Not Later? Is the Generation of TE and ICM Sufficient for Recognizing an Embryo?

The proposed definition argues that some events normally seen during human development can occur in an entity that is not a human embryo. Using similar logic, many have argued that the defining characteristics of a developing human are only evident much later, when specific developmental "landmarks" are attained—for example, implantation,[34] formation of the primitive streak,[35] "individuation" (loss of the ability to form identical twins),[36] or acquisition of higher brain functions, such as consciousness.[37] In many cases, these arguments conflate the onset of human life (as defined by some later-arising characteristics) with the onset of "meaningful human life" or human personhood. Yet in all cases, these views differ significantly from the current definition in holding that an embryo comes into existence *only* when the landmark characteristic is attained; that is, everything occurring *prior* to acquisition of the defining trait is seen as an event occurring in a "preembryo." Put another way, these arguments do not propose that later-occurring events are merely a more robust way of recognizing an embryo, but rather that they are *necessary* for an embryo to exist. The implicit conclusion is that prior to production of these later, definitive traits a "preembryo" is not a *human being*, or at least not a human "person" with moral worth, because it lacks a necessary characteristic that marks the beginning of human life (see also the discussion of human value in chapter 8 under the heading "Human Embryos and Human Value").

Thus, under the view that a functional brain is the defining land-mark for a human being to be present, if an embryo/fetus develops normally up until the twenty-fourth day and then brain formation is disrupted, such an anencephalic individual need not be considered a human being (or at least not a human being with moral worth), despite the normal maturation of other human organs that takes place prior to and after twenty-four days. Indeed, some have proposed the criteria for infant death be suspended in such cases to allow more widespread use of anencephalic newborns as organ donors.[38]

There are at least three reasons why it is both misleading and inaccurate to define an embryo based on later developmental events, with all earlier events being characteristics of a "preembryo." First, all of the proposed landmarks for what constitutes an embryo (implantation, formation of the primitive streak, "individuation," brain function, etc.) are inherently *arbitrary*. For example, the point at which various authors judge the nervous system to be sufficiently mature for a human being to be present ranges from the onset of neural function at approximately ten weeks,[39] to the onset of "coordinated neural function" at approximately twenty-three weeks,[40] to the establishment of more complex brain connections presumed to mediate "consciousness," between thirty and thirty-five weeks.[41] The variation in these figures illustrates the fatal flaw in a definition based on structural or functional maturation; unlike the clear, *qualitative* distinction between an organism and a nonorganism, embryonic development is a continuum. Consequently, there is no clear point at which "landmarks" capable of distinguishing between a preembryo and a human being are achieved. Indeed, some have argued that the brain is insufficiently mature at birth to warrant considering a newborn a human being.[42] Thus, using brain development (or any other aspect of maturing form or function) as a basis for determining what is a human being (or when human life begins) is necessarily arbitrary, and therefore a scientifically and logically inadequate definition.

Second, development reflects the *maturation* of an existing organism, not the *transformation* of a nonorganism into an organism. Just as the playing of the computer file does not "transform" a recording from a "Pre–Alphabet Song" to the actual song at the fourth measure, embryos are not transformed from a "preembryo" into a human being once some developmental event occurs. In both cases, the information intrinsic to the entity from the beginning is merely playing out over time. While later events may provide more reliable indicators that a developmental sequence is under way, they do not *produce* an embryo any more than formation of TE and ICM produces an embryo. We may have greater confidence after implantation, gastrulation, or brain formation that the entity is indeed progressing along a normal trajectory of human development, but the inherent nature of the entity as an embryo was established at its initiation, well prior to the gradual emergence of characteristic human traits.

Finally, it is important to identify the earliest reliable marker for the presence of a human organism. Once a developmental sequence is underway, all subsequent events are clearly part of that sequence. While later events may be important for the production of mature human traits, if they are part of an ongoing developmental trajectory, they are not necessary to determine that the entity is a human organism undergoing an autonomous process of self-maturation. As soon as an entity capable of development comes into existence, an embryo is present, even if an internal defect prevents it from achieving a normal, mature human state.[43] Production of the first two committed cell types in an ordered developmental sequence is the minimum requirement to establish that an entity is capable of development.

CHAPTER 4

TWINNING AND THE BEGINNING OF HUMAN LIFE

Thus far, analysis of the scientific evidence clearly indicates that a human being comes into existence at the moment of sperm-egg fusion and that the capacity to undergo development is the defining characteristic of a human being. In contrast to this view, it is often asserted that the phenomenon of identical twinning precludes the possibility that human life begins at conception and raises serious concerns regarding the nature, individuality, and moral status of the human zygote.[1] Yet are these concerns consistent with the scientific facts of twinning? Here, I will review what is known about twinning in light of the major philosophical concerns raised by this phenomenon.

How Does Twinning Occur?

Twinning can occur either because two oocytes are ovulated and fertilized, resulting in nonidentical (fraternal or dizygotic) twins, or because a single embryo splits to generate two identical (or monozygotic) twins. Yet only monozygotic twinning raises significant questions regarding human individuality and when human life begins.

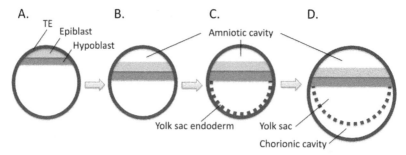

Figure 4.1. Organization of the amniotic and chorionic membranes determines the timing of twinning. (**A**) At the blastocyst stage, there is an outer layer of TE (black) and the inner cell mass (ICM), consisting of epiblast (light gray) and hypoblast (medium gray; cf. figure 2.1E). (**B**) The amniotic cavity forms between the epiblast and the TE. (**C**) Yolk sac endodermal cells (dashed line) migrate out of the hypoblast and line the blastocyst cavity. (**D**) Space opens between the yolk sac endoderm and the TE, producing the chorionic and yolk sac cavities. The amniotic and chorionic "membranes" consist of the cells lining these cavities.

Monozygotic twinning[2] is a rare event, occurring in approximately 0.35 percent of all live births.[3] The incidence of identical twinning is much higher in assisted reproductive technologies (ART), representing approximately 2 percent of all live births.[4] Although the precise mechanisms of human monozygotic twinning are unknown, it is widely[5] (albeit not universally[6]) accepted that the arrangement of the embryonic membranes (the amnion and the chorion) reflects the timing of twinning (figure 4.1; table 4.1). The blastocyst is the basis for the formation of both the amniotic and chorionic cavities, and therefore, blastocyst formation is a critical event for determining the timing of twinning.

The amnion arises after blastocyst formation when a space opens between the epiblast and the TE (figure 4.1A, B). Somewhat later, the chorion arises when a space opens between the yolk sac and the TE (figure 4.1C, D). Consequently, splitting of the embryo *prior* to blastocyst formation—either at the two-cell stage or at the morula stage (figure 2.1B, C)—produces two separate embryos, each of which will proceed to develop to the blastocyst stage (figure 4.1A) and subsequently produce

Table 4.1. Timing of twinning

Name	Acronym	Event	Timing	%
Dichorionic-Diamniotic	DCDA	Splitting of the embryo either prior to blastocyst formation or at hatching.	Day 2–5	33%
Monochorionic-Diamniotic	MCDA	ICM splitting within the blastocyst, prior to amniotic cavity formation (most likely at hatching).	Day 3–8	66%
Monochorionic-Monoamniotic	MCMA	ICM splitting within the blastocyst after amniotic cavity formation, or fusion of two amniotic cavities in MCDA twins to form a single cavity.	Day 8+	<1%

its own amniotic and chorionic cavities. Such dichorionic, diamniotic (DCDA) twins represent approximately 33 percent of all identical twins (table 4.1).

In contrast, if the ICM separates into two distinct masses *after* the formation of the blastocyst yet *prior* to the formation of the amniotic cavity (figure 4.1A), each part of the separated ICM will develop independently, forming its own mature body structures and its own amniotic cavity. In this case, the twins will have separate amniotic cavities and a shared chorionic cavity (figure 4.1). Such monochorionic, diamniotic (MCDA) twins represent 66 percent of all identical twins (table 4.1).

Finally, if the ICM separates into two distinct masses after formation of the amniotic cavity (figure 4.1B), the resulting twins will share a single amnion and a single chorion, or (potentially) be conjoined. Such monochorionic, monoamniotic (MCMA) twins are rare (less than one percent of all identical twins; table 4.1).

Different Philosophical Challenges for Different Forms of Twinning

Separation of a developing embryo at the blastocyst stage or later (MCDA and MCMA twins) clearly involves perturbation of an ongoing developmental sequence produced by a single individual that has already formed TE and ICM. Because twinning at the blastocyst stage reflects the splitting of an embryo that is manifestly undergoing a unified developmental process, it does not call into question the organismal status of the original zygote. Such twinning is comparable to asexual reproduction of any other organism that is capable of reproduction through splitting. For example, earthworms, planarian worms, and many plant species can be split, and each half will regenerate a complete organism. Clearly, when a worm is split into two worms, the ontological status of the original worm as a single, unified organism is not called into question. Yet MCDA or MCMA twinning does raise a number of important philosophical issues that will be addressed below, including the following: Does the original embryo die or continue as one of the twins? What is the moral worth of the embryo prior to twinning? Who are the parents of the twins?

In contrast, if twinning is due to totipotent cells present in the early embryo establishing independent developmental trajectories, it is often argued that no single embryo can exist until such totipotent cells no longer exist. The standard interpretation of the timing of twinning suggests that separation of early blastomeres is not a rare event (i.e., DCDA twins are 33 percent of all identical twins), and this interpretation also heightens the concern that if a zygote can give rise to more than one mature individual, the zygote cannot be seen as a single individual. If this is the case, it raises a number of serious issues, including the following: What is the moral worth of the zygote? How many totipotent cells can coexist in a single embryo? When does human life actually begin?

Is the Standard View of Twinning Correct?

DCDA twinning, particularly at the two-cell stage, raises the most serious philosophical issues, yet it is important to ask: How likely is it that embryos ever actually split at the two-cell stage? The standard view of twinning can be challenged on a number of fronts. Most notably, while twins can be experimentally produced by splitting at the two-cell stage,[7] there is no direct evidence that this ever naturally occurs.

The first days of life cannot be observed in natural conception, but they are routinely examined during the use of assisted reproductive technologies (ART), where the incidence of monozygotic (MZ) twinning is also significantly elevated. Yet despite the daily observation of ART-produced embryos, twinning in the first three days of life has never been reported. Indeed, one author states, "We have never observed an embryo spontaneously splitting in half before the blastocyst stage in over thirty years of laboratory experience."[8]

In addition to this negative evidence, four positive lines of evidence indicate early embryo splitting is unlikely. First (as noted above), when blastomeres from freshly dissociated morula-stage embryos are reaggregated, they reconstitute an intact embryo that proceeds through development to live birth,[9] indicating that cells of the early embryo are both highly adherent to each other and strongly inclined to act as an integrated whole.

Similarly, when blastomeres from two or more embryos are aggregated, they combine to form a single, chimeric embryo[10]—that is, an embryo made up of a mixture of cells with distinct genetic identities. This is true even when the blastomeres are derived from different species; for example, aggregated sheep and goat blastomeres that are transferred to a surrogate mother produce live-born, chimeric "geeps" consisting of a mixture of sheep and goat cells.[11]

Third, embryonic stem cells (ESCs) have properties very similar (albeit not identical[12]) to those of early blastomeres. When ESCs are injected into intact embryos, they do not remain as an independent, autonomously developing cell mass, but instead rapidly integrate into the ICM and participate in the ongoing embryonic trajectory,[13] again indicating the propensity of embryo-derived cells to both adhere to each other and act as parts of a unified whole.

Finally, the possibility that two early embryos could develop independently in close proximity to each other has been directly tested. Early embryonic development takes place within an acellular protein coat

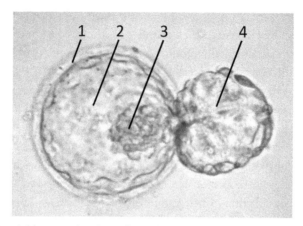

Figure 4.2. A blastocyst hatching from the zona pellucida. (**1**) The zona pellucida. (**2**) The portion of the embryo still within the zona. (**3**) The inner cell mass. (**4**) The portion of the embryo outside the zona. Image modified from "Human blastocyst hatching," by K. Hardy. Credit: K. Hardy. CC BY. Available at the website of the Wellcome Collection, https://wellcomecollection.org/works/scfp2gf2 ?query=hatching&page=1.

known as the zona pellucida (figure 2.1; figure 4.2). In a recent study, over one hundred immature mouse oocytes were manipulated to artificially produce two mature oocytes within a single zona. These "binovular" oocytes were fertilized, generating fraternal twin-zygotes, also contained within a single zona. In all cases, the twin embryos rapidly annealed to form a single, chimeric embryo.[14] This result strongly suggests that even if an embryo were to split while still within the zona pellucida, the resulting twin embryos would rapidly reanneal to reform a single embryo.

All of these lines of evidence strongly suggest that twinning at the two-cell stage is highly unlikely due to the inherent propensity of early blastomeres to adhere to each other and to collectively participate in a single, unified embryonic trajectory. Yet in light of this evidence, how do the 33 percent of monozygotic twins with separate chorionic and amniotic cavities (DCDA) actually arise?

One possibility is that twins form upon emergence of the embryo from the zona pellucida (figure 4.2). The function of the zona is to protect the embryo from damage as it travels down the fallopian tube to the uterus. However, the outer cells of the embryo (the trophectoderm or TE) are responsible for interacting with the uterine lining and mediating implantation, an event that is crucial for the survival of the embryo. For implantation to occur, the embryo must escape from the zona in a process known as "hatching" (figure 4.2), so that the cells of the TE are free to attach to the uterus. This happens by an active process initiated by the embryo.[15] Cells of the TE secrete enzymes that degrade the zona, weakening it so that the embryo is able to squeeze out of a small hole (figure 4.2). During extrusion of the blastocyst from the zona, the cells of the ICM can be separated into two independent groups of cells, resulting in the formation of MCDA twins (two separate ICMs within a single blastocyst).[16] Alternatively, the entire embryo can split in two, giving rise to DCDA twins (two separate blastocysts).[17]

The incidence of human monozygotic twinning is increased approximately sixfold (to 2 percent) during ART procedures,[18] and in mice, the incidence of ICM splitting at hatching is also greatly increased when embryos are cultured.[19] Interestingly, several studies of ART suggest there is a higher incidence of monozygotic twinning with blastocyst transfer (day five) than with early embryo transfer (day three),[20] suggesting that longer times in culture may also increase the likelihood of blastocyst splitting in humans. Finally, there have been at least two clinical observations of human twinning through blastocyst splitting at hatching.[21] Since most monozygotic twinning observed during ART procedures occurs following transfer of a single blastocyst to the uterus, it is likely that twinning occurs through splitting of the blastocyst following transfer. Splitting at the blastocyst stage may also underlie naturally occurring DCDA twinning, albeit at a lower frequency than is seen in ART procedures.

In summary, 66 percent of identical twins (table 4.1, MCDA and MCMA) are likely to occur by splitting of the ICM after blastocyst formation. This (1) indicates that a single developing embryo (as evidenced by the production of both TE and ICM) has split into two and (2) does not call into question the ontological status of the original embryo. While the remaining 33 percent of DCDA twins could theoretically occur by separation of the first two blastomeres, there is no evidence that this naturally occurs and ample evidence suggesting that if it were to occur, the twins would rapidly reanneal to form a single embryo. It seems far more likely that DCDA twins arise by blastocyst splitting at hatching, an event that has been observed during ART procedures and one that also does not call the ontological status of the original embryo into question.

The likely timing of monozygotic twinning, combined with the immediate onset of organismal function following sperm-egg fusion, allows us to identify a seventh important unifying concept in consideration of the human embryo:

UNIFYING CONCEPT 7: There is clear scientific evidence that the one-cell embryo or zygote initiates a developmental trajectory; that is, the zygote is manifestly a human organism. Therefore, twining at the two-cell stage or later does not call into question the ontological status of the original embryo as a complete and *individual* human being.

CHAPTER 5

PHILOSOPHICAL CONCERNS REGARDING TWINNING

The most serious concerns raised by twinning involve splitting of the embryo at the two-cell stage. As we have seen, this is highly unlikely. However, if it is theoretically possible, regardless of how infrequent, it must be addressed. For this discussion, I will assume twinning by separation of the first two cells of the embryo is "possible," even if it never actually occurs. Splitting at later stages does not call the beginning of life or the nature of the zygote into question, but raises concerns about individuality. The philosophical and logical issues raised by embryo twinning have been discussed in detail elsewhere.[1] Here, I will briefly address the major concerns, in light of the scientific facts regarding when human life begins and the biological mechanisms of monozygotic twinning.

Developmental Potential and "Virtual" Humans

The most fundamental philosophical concern is whether the phenomenon of monozygotic twinning precludes the zygote from being an individual human on purely logical (and therefore also, ontological) grounds. This argument was initially put forward in 1970 by Joseph Donceel,[2] and was expanded by Norman Ford in 1988.[3] Numerous modern authors have also taken up this argument in various forms.[4] In brief, Donceel argues that while a zygote has the *potential* to develop into a mature human, it is not a human individual until the possibility of twinning has passed. This is the case, Donceel argues, because if "potential" to develop into a mature human is the criterion for being a human now, then until totipotency is lost, "every single cell of the zygote, of the morula, or of the blastula, is a human person."[5]

A similar objection is voiced by Ford, who acknowledges that the zygote is a living individual but questions whether it is a living, individual, human *person*. Like Donceel, Ford argues that although the zygote has a "natural active potential" for development, it cannot be an individual human being because

> once we assume that the zygote is a human individual because it has the natural active potential to develop into an adult we begin to run into difficulties. The same zygote would also have the natural active potential to develop into two human individuals by the same criteria. We could legitimately ask whether the zygote itself would be one or two human individuals. It would seem absurd to suggest that at the same time it could both be one and more than one human individual, granted that each must be a distinct ontological individual.[6]

Thus, both Donceel and Ford conclude that until totipotency is lost, the embryo can at best be considered a "virtual" human—that is, a cell that has the potential to become a human but is not an individual human yet.

Some modifications of this argument are warranted in light of the scientific facts. First, our understanding of embryology has advanced considerably since 1970, and we now appreciate that totipotency is unlikely to be preserved beyond the two-cell stage. Therefore, in most cases only one developmental trajectory is initiated by the zygote, and in the worst case, only two;[7] that is, it is never the case for a blastocyst-stage embryo containing many hundreds of cells that every cell is totipotent and that therefore "every single cell . . . is a person."

Second, ample scientific evidence indicates quite unambiguously that the life of a human organism begins at sperm-egg fusion.[8] Therefore, when twinning occurs, whether at the two-cell stage or at the blastocyst stage, a single, developing organism is split. This observation largely defuses the ontological question of origin since, just as an earthworm is an individual organism prior to splitting, the zygote is an individual human being prior to twinning.

However, the central issue in the argument of both Donceel and Ford is one of "potency." If a zygote is a person by virtue of the fact that it has the potential to develop into a mature human, then all cells with this potential (i.e., each individual blastomere that can exhibit totipotency when isolated, be there hundreds or only two) are equally persons. Moreover, a zygote that has the "potential" to be two individuals cannot simultaneously be considered a single individual.

The key to resolving this dilemma is a more precise understanding of the term "potency." Aristotle takes up the different meanings of "potency" in *De anima* (2.5; a general discussion of potency may be found in *Metaphysics* 5.12) and defines this term along two distinct axes: active-passive and proximate-remote.

Active potency is a power to act in a specific manner that is intrinsic to the thing itself; for example, I have the active potency to speech, even if I am currently silent, while a tree does not. *Passive potency*, in contrast, is a power to be acted *upon* in a specific manner; for example, wood has the passive potency to be fashioned into a chair by the action of a carpenter, while water does not. And neither wood nor water has the *active* potency to become a chair on its own.

Similarly, *proximate potency* is a power to act or to be acted upon in a specific manner that can be immediately exercised; for instance, I have the proximate potency to speech, even if I am currently silent. Lumber has the proximate potency to be acted upon by a carpenter, even if it is currently sitting in a lumberyard. In contrast, *remote potency* is a power to act or to be acted upon in a specific manner that requires some intervening event before it can be exercised; for example, an infant has the remote potency to speak, but must first learn a language. A seedling has the remote potency to be fashioned into a chair, but must grow to be a tree first.

Applying these distinctions to the embryo, it is clear that the zygote exhibits a *proximate, active potency* to develop as a human being; that is, a zygote is a human *now* because it is actively exercising an intrinsic power of maturation/development.

In contrast, early blastomeres that can develop independently when isolated from the embryo exhibit only a *remote active potency* to be a human being while they are still part of an intact embryo. Blastomeres are parts of a single human now, but can become independent human beings if an intervening event (i.e., isolation of the blastomere) occurs. Importantly, the remote potency of blastomeres means that there is a single, unified human individual prior to twinning, and that blastomeres must cease being a *part* to actualize their (remote) potential to be a *whole* on their own; that is, they must undergo a substantial change.[9]

Embryonic stem cells (ESCs) can be contrasted to both zygotes and early blastomeres. ESCs that are injected into a developing embryo can participate in embryonic development, forming part (or in some cases, most) of the mature body. Yet unlike zygotes and blastomeres, which can develop independently when isolated, stem cells exhibit only a *passive, re-mote potency* to be part of a human being; that is, stem cells are not parts of a human now and cannot become parts of a human on their own, but can be acted upon by an embryo (or an experimenter) to be converted into a *part* of a developing individual.[10]

Thus, Donceel and Ford are correct that merely having some kind of "potential" to develop to mature stages of human life cannot be the criterion for actually being an individual human person, but they do not adequately distinguish between different types of potency. *Proximate active potential* to develop to mature stages of human life is the criterion for being a human individual now. This potential exists in a zygote and is a property of the embryo as a whole at later stages of development, but does not exist in blastomeres that are part of an existing embryo (yet can develop independently when isolated) or in stem cells.[11]

Blastomeres, regardless of their potency, are merely *parts* of a human being. Upon splitting/twinning, the remote, active potential of the blas-tomeres to develop as a full human individual becomes proximate (i.e., a substantial change from a part to a whole occurs), and at that point, two human beings exist. Theologically, the transition from a part to a whole would also require a new human substantial form (or soul) to come into being at the instant of splitting. In contrast to the view of Ford,[12] this would not be a case of "delayed hominization," since the substantial form of the original embryo persists after twinning (see below) and the ensoulment of the newly generated embryo occurs immediately upon formation of the twin.

> UNIFYING CONCEPT 8: Proximate active potential to develop to mature stages of human life is the criterion for being a human organism—that is, a human being at the beginning of the human life span.

No Corpse

A second concern regarding twinning involves what happens to the original embryo after the twinning event. While Ford acknowledges that twinning involves a substantial change (the change from a single "individual" to two "individuals"), he finds it "paradoxical, but still necessary, to admit that the original zygote and human individual cease to exist, when, without dying and without a dead cell remaining, it gives asexual origin to identical twin offspring."[13] A similar concern is voiced by Anscombe, who asks, "But what has become of the human that both of them [the twins] once were identical with? Has he—or it—simply ceased to exist, as we might say the parent amoeba ceases to exist on splitting?"[14]

Understandably, the notion that a new individual or individuals can be generated from the zygote without the zygote itself dying appears contradictory. Yet there are many examples of substantial change occurring without the requirement for a corpse. As Anscombe notes, cell division (or mitosis) routinely produces two cells out of one cell without a corpse. In the converse direction, sperm-egg fusion produces a single zygote out of two living cells without the gametes "dying." Importantly, substantial change does not require a death, but merely a ceasing to be; apt matter persists and receives the new form. In the case of twinning, I will argue below that the original embryo persists without ceasing to be and that a *part* of the initial embryo undergoes a substantial change to become a complete individual (with the part ceasing to be, but not "dying," since living cells persist throughout the process of twinning).

Ford and Anscombe are correct in insisting that there must be *evidence* for a new form coming into existence. The evidence for a substantial change converting human gametes into a human being upon fusion is the initiation of *development*; that is, the newly formed zygote immediately produces a molecular cascade that controls its subsequent maturation as a human organism. Similarly, the evidence for two embryos coming to be as a consequence of twinning is that each half of the split embryo assumes an *independent* developmental trajectory.

As an illustration of what occurs following splitting, consider the simplified case where a blastocyst-stage embryo is split and one half is discarded (figure 5.1). The resulting "demi-embryo" rapidly reseals and proceeds immediately with development,[15] maturing in synchrony with unsplit sibling embryos.[16] Direct observation indicates that the ratio of cells in the ICM and TE of the demi-embryo is either maintained or restored by cell proliferation,[17] with lineage analysis indicating that the TE and ICM largely contribute cells to their own lineage, just as they do in normal development.[18] Thus, without a perceptible pause, the surviving half immediately proceeds as a unified whole along the developmental trajectory that was established by the zygote and was ongoing prior to splitting. While the original embryo has clearly sustained a serious injury, there is no evidence that it has ceased to be.

What occurs after splitting most closely resembles the process of organismal "regulation" or wound healing, first described by Driesch in 1892.[19] Just as a mature human organism that has been injured continues to be a human (albeit a damaged one), a blastocyst-stage human embryo that has lost half of its cells is still a *human organism*. And one of the characteristics of organisms is that they *repair* injuries—which is precisely what occurs following splitting. The fact that an embryo is more adept at repair and regeneration than an adult does not preclude the embryo from being a human individual. Therefore, in the example given above, it is reasonable to view the surviving demi-embryo as an injured human organism who is ontologically identical to the initial embryo.

Logically, the ontological status of the surviving half does not depend on the (as yet undetermined) status of the second (discarded) half. Yet if the embryo is split in half and both halves survive and regenerate their missing parts, there is still no evidence that the initial embryo ceases

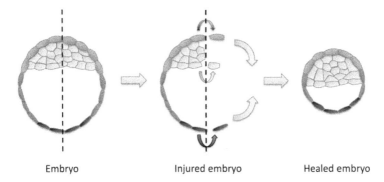

Embryo   Injured embryo   Healed embryo

Figure 5.1. Twinning by embryo splitting at the blastocyst stage. Cells of the blastocyst have distinct molecular properties and developmental functions. After splitting (dashed line), a closed sphere rapidly reforms (large curved arrows), and cells are replaced within their own tissues (small curved arrows); that is, trophectoderm (dark gray) regenerates trophectoderm and ICM (light gray) regenerates ICM. There is no evidence for respecification across lineages. The smaller, "demi-embryo" has approximately half as many cells as the original blastocyst and develops from the blastocyst stage in synchrony with unsplit sibling controls. An earlier version of this diagram appeared in M. L. Condic, "Totipotency: What It Is and What It Is Not," *Stem Cells and Development* 23, no. 8 (2014): 796–812.

to be, indicating that one of the twins is ontologically identical to the original embryo, and one is a newly generated individual. Importantly, our inability to tell the twins apart is an epistemological, not an ontological, problem.[20] We may not know with certainty which half is the original embryo, but there is clearly no evidence the original embryo has ceased to be.[21]

> UNIFYING CONCEPT 9: Monozygotic twinning most closely resembles regeneration or wound healing of an injured human organism; development proceeds without pause along the trajectory initially established by the zygote, indicating that one of the two twins produced by splitting is ontologically identical to the original zygote.

Thus, Ford and Anscombe are rightly concerned that no new individual can come into existence without the matter constituting that individual undergoing a substantial change, but this does not require a corpse. In the case of twinning, the half of the original embryo that becomes the new individual undergoes a substantial change from a *part* to a *whole*,[22] a change that involves *transformation* of cells that were once part of an embryo into a unified organism, but not death.

Unnatural Reproduction: Who's Your Daddy?

A third concern raised by twinning is that of "unnatural" reproduction. For example, in a 1994 article, William Wallace objects:

> If God created the human soul and infused it into the zygote at the moment of fertilization, then a stable individual of human nature would already have been formed. And, were another individual to be formed subsequent to that moment, this would be an instance of asexual generation, the production of one individual from another of the species—a type of generation found in lower forms but not proper to humans.[23]

Asexual human reproduction is certainly *unfamiliar* to most of us; we don't routinely see humans split in two or "bud" to produce a new individual. Yet from a scientific perspective, it is simply a factual observation that for the first fourteen days of life, asexual reproduction by splitting (i.e., identical twinning) is a rare, yet completely natural, form of human generation.

Setting aside the unfamiliarity of asexual reproduction, it seems that underlying Wallace's objection is a concern regarding parenting and the relationship of the twins to each other. In Wallace's view, if twinning initiates with an actual human individual (the zygote), then the original zygote is both the brother (or sister) and the father (or mother) of the twin that is generated. And to Wallace, this appears unseemly.

This concern can be addressed by drawing a clear distinction between random and intentional acts. Aquinas defines intention in two ways: primarily as the rational ordering of an act to a specific end and secondarily as the *natural* ordering of acts. In defense of this second sense, Aquinas states (*Summa Theologica* I-II, q. 12, a. 5, sed contra):

> To intend is to tend toward something, which is indeed said of the mover and the thing moved. Accordingly, that which is moved to an end by another is said to intend the end. In this way, nature is said to intend an end, as it is moved to its end by God, just as the arrow is moved by the archer. And in this way also brute animals intend an end, in as much as they are moved by natural instinct to something.

Importantly, intention does not require "choosing." In Aquinas's view, an arrow shot by an archer *intends* to hit its mark. Swallows, driven by instinct, *intend* to return to Capistrano. Neither makes a rational choice to move in a particular direction.

Equally importantly, some acts, such as authorship and artistic creation, can *only* be intentional, because they reflect rational human agency. If a monkey at a typewriter randomly produces the text of Shakespeare's Hamlet, the monkey cannot be legitimately named as the author of this work, because the act of randomly striking typewriter keys is not naturally or rationally ordered to any specific literary end. Similarly, if the growth of a tree randomly produces an image of the Virgin in the bark, the tree cannot be legitimately named as the artist creating this image, because the act of growth is not naturally or rationally ordered to the production of religious images. However, acts of living beings that are inherently determined by nature (e.g., the swallows returning to Capistrano) can *also* only be intentional, because "nature is said to intend an end, as it is moved to its end by God." In the same manner, the actions of cells and molecules within a living organism that are ordered ("moved") by an intrinsic organizing principle (i.e., a substantial form) can only be considered intentional. Thus, binding of oxygen by hemoglobin is "intentional," in that it reflects the inherent nature of the hemoglobin molecule and its ordered function within a red blood cell of a living being.

In a similar manner, parenting can *only* be intentional, because (independent of the conscious intentions of the human actors), reproductive acts are inherently ordered towards reproduction. The terms "mother" and "father" refer to those agents who participate in a sexual reproductive act as *reproductive*. Whether through natural instinct (as is the case for animals) or through conscious intention (as is often the case for humans), whether out of love of pleasure, or love of each other, or even out of violence (as in the case of rape[24]), "mother" and "father" refer to those agents who act *for the end of reproduction*—that is, who act intentionally.

Yet, as we have seen, when an embryo secretes enzymes to degrade the zona pellucida, this act is naturally ordered towards *implantation*, not *reproduction*. Consequently, if twinning occurs as a random consequence of embryo hatching, due either to some atypical feature of the zygote itself or to some external agent (such as an abnormal intrauterine chemistry), it is still the father and mother of the original zygote, as the intentional originators of the reproductive act, who are the parents of the twin.[25]

Thus, Wallace is correctly concerned that human individuals cannot be generated without the involvement of human parents, yet the parents of the original zygote, as the intentional originators of the reproductive act, are also the parents of the twin.

> UNIFYING CONCEPT 10: Sexual intercourse is intrinsically (or naturally) ordered to reproduction and therefore parenting of a child can only be an intentional act. In contrast, monozygotic twinning is a random disruption of actions that are naturally ordered toward development, not reproduction. Therefore, in cases of monozygotic twinning, the parents of the original embryo are the parents of the twins.

Identity in Form and Matter versus Ontological Identity

A component of all of the philosophical concerns raised by Donceel, Ford, Anscombe, and others about twinning is the question of identity in form and matter. The transitive nature of identity requires that if A equals B, and A equals C, then B must also equal C. Applying this principle to the embryo, if the original zygote is identical to both of the totipotent daughter blastomeres that will (when separated) develop as twins, then the twins are identical to each other, which is clearly not the case. Formally, the problem of transitivity proceeds as follows:

Zygote A is identical with blastomere B, *and*
zygote A is identical with blastomere C,
but two things identical with some third thing are identical with
    each other.
∴ Blastomere B is identical with blastomere C.

Some modifications to the concerns regarding transitivity are warranted in light of the scientific facts. As we have seen, twinning is unlikely to occur at the two-cell stage, and substantial evidence suggests that in most cases, the first two blastomeres are not totipotent and therefore not identical. Moreover, substantial evidence also indicates identical twins are epigenetically distinct, even at early stages.[26] Consequently, zygote A is unlikely to be "identical" to both of the daughter blastomeres, although it may be "nearly identical."[27]

There are three related responses to the question of the ontological identity of twins; (1) identity (or near identity) is not "sameness," (2) structural-functional identity does not confer ontological identity, and (3) there is an important difference between traits used to fix a name to an individual and that individual's identity.

(1) There are many cases in which two entities that are identical or nearly identical are nonetheless not the same in essence or in kind. For example, because a mature oocyte is roughly ten thousand times larger than a mature sperm,[28] an oocyte and a newly-formed zygote are more than 99.99 percent identical in their molecular composition. Yet there is a clear ontological difference between an oocyte (a human cell) and a zygote (a human being).

Similarly, whole genome sequencing has determined that the human and chimpanzee genomes are 98 percent identical,[29] and yet few would seriously argue that humans and chimpanzees are ontologically identical.

Finally, although the instant of human death is difficult (if not impossible) to determine with precision, death clearly does not happen by degrees; that is, humans are either alive or dead, they are never "half-dead." However, death does not occur because some physical substance leaves the body. Consequently, immediately prior to death and in the instant following this event, the composition of the body is identical. Yet despite the *identical composition* of a living human being and a freshly formed corpse, the two are ontologically distinct.

(2) Donceel, Anscombe, Ford, and others appear to be using a confused notion of identity.

All of these authors acknowledge that monozygotic twins (believed to have arisen from "identical blastomeres") are ontologically distinct, even though they are presumed to be structurally and functionally identical. Why is this not also the case for a zygote and the (presumed) "identical" blastomeres it generates? Ford acknowledges that in cases where separated blastomeres mature into twins, they both have "separate concrete existences,"[30] an acknowledgment that logically requires the two entities (regardless of any similarity in their composition) to also be ontologically distinct—that is, separate members of the same species.

(3) Saul Kripke provides an important insight into this problem by drawing a distinction between naming and identity. Kripke notes that the attributes by which we fix a name to an individual do not make the individual be who he is.[31] In more precise terms: The substantial identity of a thing is logically prior to the attributes distinguishing that thing from anything else. To use Kripke's example, we may know Benjamin Franklin to be Franklin by virtue of a specific attribute; for example, Franklin is the inventor of bifocals. Similarly, we may know Thomas Jefferson to be Jefferson by a distinct attribute; for instance, Jefferson is the author of the Declaration of Independence. However, if we are mistaken, and it turns out that Jefferson, not Franklin, actually invented bifocals, this does not turn Jefferson into Franklin. An attribute used to name (or distinguish) one human from another does not "make" them distinct.

Applying Kripke's insight into the case of twinning by separation of two hypothetically "identical" blastomeres, even if there are no unique attributes distinguishing the blastomeres, *they are not made ontologically identical by this fact.* It may be difficult (or impossible) to distinguish the two blastomeres (i.e., fix a name to them), but they still have "separate, concrete existences," which requires them to have independent substantial forms. As noted earlier, the difficulty of determining which twin is which (i.e., fixing a name) is an epistemological, not an ontological, problem.

Thus, Donceel and others correctly insist that the transitive nature of identity requires that two individuals who are identical to a third individual also be identical to each other, but this fact does not include the substantial identity (or substantial form) of the individual, which is logically prior to any attributes that might distinguish one human from another.

> **UNIFYING CONCEPT 11:** Monozygotic twins are unlikely to be truly "identical." Distinguishing them from each other and determining which twin is ontologically identical to the original zygote is an epistemological question. Importantly, human individuals are not made to be distinct by distinguishing differences between them.

Alternative Possible Futures

A final concern raised regarding twinning is that of alternative possible futures, an argument most clearly articulated by Berit Brogaard.[32] Brogaard considers three possible mechanisms of twinning: budding (separation of a group of cells from an initial organism, with the group becoming an independent organism), fission (splitting of a single cell into two identical cells, with concomitant destruction of the initial cell), and separation (or division of two cells or groups of cells into two separate organisms).

Brogaard rejects budding because, "unfortunately, . . . human development is nothing like that of plants. When a cutting is taken from a fully developed plant, we have an original organism and a part that develops into a separate individual. We do not have one cell or mass of cells, which divides in two."[33] Yet as we have seen, human twinning is quite similar to budding. For twinning at the blastocyst stage, the most reasonable interpretation is that one organism/twin continues to develop along the same trajectory as did the initial embryo and a new organism is produced from a separated part, just as in plants. The only difference between this case and that of plants is the relative size and maturity of the two organisms, a difference that does not introduce a meaningful philosophic complication.

Brogaard also rejects fission, both because it raises the problem of transitivity discussed above and because, "as it turns out, . . . human embryos do not undergo twinning via fission. The reason is that the cells—and sums of cells—in the pre-gastrula embryo are totipotent: each of them has the potential to develop into a complete human being."[34] As discussed, it is not the case that all cells of the pre-gastrula are totipotent. Yet if twinning occurs by division of the zygote to generate two blastomeres and the subsequent separation of these first two cells to yield two totipotent zygotes, it would be *exactly* like fission, raising the same questions of transitivity discussed above.

Brogaard favors separation as the mechanism of twinning, which also raises the problem of transitivity, albeit in a modified form. Brogaard argues that if an existing embryo composed of parts A and B is transtemporally identical to a single born individual C, yet the parts A and B could, in an alternative possible future, give rise to two distinct individuals through twinning, then the born individual C is identical to both of the possible individuals the parts could have produced. Formally, the argument proceeds like this:

Born human C = embryo composed of A+B;
embryo composed of A+B = possible born human A and possible born human B;
but two things identical with a third thing are identical with each other.
∴ Born human C is identical with possible born human A and possible born human B.

Brogaard rejects the possibility that the initial embryo is a whole and that separation of A and B destroys this whole and produces two new entities, stating, "The problem with this proposal is that the cells in the early embryo form a mere mass, being kept together spatially by an outer membrane. There is no causal interaction between the cells."[35] Yet, as we have seen, this characterization is factually incorrect; the cells of the early embryo rapidly become specialized and interact with each other from the very beginning in an integrated manner to advance the development and maturation of the embryo as a whole. Consequently, the embryo prior to splitting is most accurately seen as a whole, not as an aggregate of individual parts. Splitting of this single individual could result in either the destruction of the initial whole and production of two new individuals or (as argued above) the continued existence of the original individual and the concomitant production of a new individual.

The more serious problem with Brogaard's argument arises from equating *possible futures* with *actual futures*, without adequately accounting for the difference between proximate and remote potency. While a single embryo composed of parts A and B is in remote potency to the "possible" future of producing two distinct individuals (should splitting occur), in actuality, only a single developmental trajectory is operative within the embryo prior to splitting.

To use the more familiar example raised by Brogaard, no one would doubt that a plant is a single, integrated organism that could, nonetheless, become multiple "possible" plants in an alternative future, following splitting. Yet, should a plant be split into two, three, or even a dozen separate divisions, all of which produce mature plants, the principle of transitivity does not require the original plant to be transtemporally identical to both its "possible" more mature self (had splitting not occurred) and also to the "possible" multitude of individual plants it *could* produce if splitting *did* occur. Splitting enables the remote potency of a *part* of the plant to become a proximate potency, resulting in a substantial change within the separated part such that it now functions as an independent whole. The new plants thus constituted are not identical to the initial plant, individually or in aggregate. They are distinct organisms.

The same holds for embryos. If an embryo composed of parts A and B would normally mature into a single individual C, then following twinning by separation, the evidence suggests the most likely case is that individual C persists, and the separated part becomes a new individual, D. The initial embryo is never identical to the new individual D and is always identical with the individual C. As Brogaard notes, distinguishing the twins may be difficult[36] because "there is no property in virtue of which one but not the other twin could be said to be identical to the ancestor entity."[37] Yet this is an epistemological, not an ontological, problem.

Thus while Brogaard is correct in pointing out that it is logically impossible for one embryo to be identical to both a single individual and two possible individuals in an alternative future, equating possible futures with an existing developmental trajectory does not adequately account for the difference between proximate and remote potency.

CHAPTER 6

PHILOSOPHICAL CONCERNS RAISED BY HUMAN CHIMERISM

How Do Human Chimeras and Mosaics Arise?

The converse of embryo splitting is embryo fusion, or the production of a single, chimeric individual from separate individuals. A chimera, or an organism that contains cells with two or more distinct genomes, can be formed in a number of ways. As noted earlier, chimeras containing cells from different animal species can be produced in the laboratory.[1] Naturally occurring chimeras arise during gestation of fraternal siblings (twins or higher-order multiples).[2] For example, if two eggs are fertilized to generate male and female twins, and these embryos (or some of their cells) subsequently combine, they will produce a single individual containing both female and male cells (figure 6.1A).

In contrast to chimeras, mosaics are individuals that contain cells with different genetic variations of the *same genome*. Typically, this is due to a genetic change that occurs in one cell of an early embryo that is passed along to all the progeny of that cell. Yet some relatively simple genetic changes can have a profound impact on the individual as a whole. For example, if one cell of a two-cell male embryo duplicates its chromosomes normally in preparation for division, yet erroneously

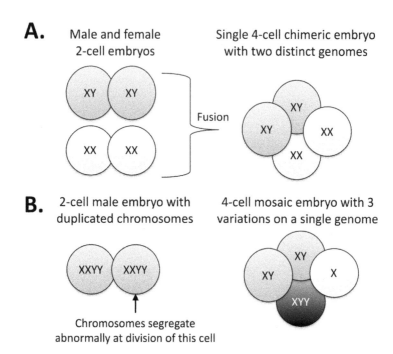

Figure 6.1. Formation of chimeras and mosaics. **(A)** A chimeric embryo can form when a two-cell male embryo (light gray) and a two-cell female embryo (white) fuse to form a four-cell chimeric embryo containing half male cells and half female cells. **(B)** A mosaic embryo can form if one cell of a two-cell male embryo divides abnormally and donates both of the duplicated Y chromosomes to one of its offspring. The resulting four-cell embryo will consist of half normal male cells (light gray), a quarter abnormal male cells with two Y chromosomes (dark gray) and a quarter abnormal female cells with only one X chromosome.

donates both Y-chromosomes to only one of its progeny (figure 6.1B, dark gray cell), the resulting four-cell embryo will contain a mixture of three distinct genotypes; 50 percent normal male cells (46XY; figure 6.1B, light gray cells), 25 percent abnormal male cells with Jacobs syndrome (47XYY; figure 6.1B, dark gray cell) and 25 percent abnormal *female cells* with Turner's syndrome (45X; figure 6.1B, white cell). In this case, the majority of the genome is shared by all the cells, but specific cells (and their progeny) will have distinct properties and distinct genetic sex.

Formation of Chimeras by Cell Incorporation

While the unusual case of intersex mosaic individuals described above presents challenging medical and gender issues,[3] chimeras (both mixed-sex and same-sex) raise the further ontological question of how two individuals can combine to form a single individual. Although the precise mechanisms underlying formation of singleton human chimeras (in contrast to chimeric twins[4]) are not known, there are at least three possibilities, all of which involve fraternal sibling embryos: (1) incorporation of cells from one twin into another (e.g., by fusion of the placental circulation), followed by death of one of the twins; (2) death of one twin, with survival of some of its cells and subsequent incorporation of those cells into the surviving twin; (3) fusion of two (or more) embryos to form a single embryo (figure 6.1A).

The first two possibilities are the least problematic. Incorporation of cells from one individual into another is relatively common during prenatal life. For example, most of us exhibit some degree of "microchimerism," due to exchange of a small number of cells between mother and child during gestation.[5] Indeed, individuals who are not their mother's first child can contain cells not only from their mother but also from all of their older siblings.[6] Similarly, fraternal twins can share cells due to fusion of the placental circulations allowing stem cells to pass between the twins.[7] Clearly, such cases of cell-incorporation do not raise significant ontological issues regarding the identity of the chimeric individuals, any more than blood donation or organ transplant calls into question the identity of the patient receiving the transplant. Incorporated donor organs (or donor cells) participate in the host's ongoing integrated, organismal function, whether the donor remains alive (as in the case of blood donation) or is already deceased (as in the case of cadaveric organ donation).[8]

Philosophically, the nature of chimeras formed by incorporation of cells turns on the distinction between a whole human and *a part* of a human. Cells or organs isolated from a living human are not organized to constitute a whole individual (i.e., they are not human *organisms*),[9] but rather they are merely *parts* of a human organism. Upon incorporation into a new individual, these parts will contribute to the ongoing organized function of that individual; that is, they undergo a substantial change and are now governed by the substantial form of the host. In some cases, the host may acquire new characteristics, based on the genetic makeup of the incorporated cells. For example, if the donated cells carry the gene for sickle-cell anemia and these cells make substantial contributions to the bone marrow, the host is likely to acquire this genetic disease. Similarly, a fetus can incorporate cells from the mother that result in the development of chronic autoimmune disease.[10] However, acquiring a disease as a consequence of cell incorporation does not call into question the ontological identity of the individual; the host is the same individual, but now with an unfortunate medical condition.[11]

There are, however, troubling cases in which cell incorporation is associated with a significant shift in the developmental trajectory of the embryo. For example, a recent review of fraternal twins who share a single chorionic cavity (dizygotic-monochorionic, or DZ-MC twins) indicates that 90 percent have some degree of chimerism, and in 15 percent of mixed-sex DZ-MC twins, one or both have a genital anomaly.[12] Such cases are most likely due to hormones passing across the fused circulatory systems, but could, at least in part, be due to hormones produced by the incorporated cells themselves. Although the number of such cases is small (thirty-one DZ-MC twins in total were reported in this study), they call into question the identity of such chimeric embryos. For example, can an individual who was developing as a normal female, yet diverts to an abnormal, intersex trajectory after incorporating cells from her twin brother, still be considered the same person?

Again, the answer to this question turns on the difference between a whole and a part. In such cases, the overall developmental trajectory of the individual persists, as does the internal developmental "program."[13] Just as incorporating cells that carry the sickle-cell gene can cause malfunction of a part (the hematopoietic system), incorporating opposite-sex cells can also disrupt the development of a part (the reproductive system). Thus, the intersex disorders observed in chimeric mixed-sex twins may be an unfortunate consequence of cell incorporation but do not alter the ontological identity of the host individual.

UNIFYING CONCEPT 12: Similar to mature individuals who receive a blood transfusion or an organ donation, chimeric individuals who are produced by incorporation of cells from a twin (whether the twin survives or dies) remain ontologically the same individual before and after incorporation. The incorporated cells are governed by the unifying principle (or substantial form) of the individual in whom they reside.

Do Chimeras Form by Embryo Fusion?

Fusion of two embryos (the final possibility noted above) is significantly more problematic than chimera formation by cell incorporation, primarily because fusion raises philosophical questions regarding the identity of the resulting embryo and the fate of the two original embryos. It is important to point out that it is currently unknown (and is perhaps unknowable) whether fusion of twin embryos ever naturally occurs; that is, in theory, *all cases of singleton chimeras* could be explained by incorporation of cells from a twin who had previously died or from a living twin who died following transfer of cells to their sibling. Neither of these two scenarios raises significant issues of identity for the surviving twin. It is only in cases where two living embryos fuse to give rise to a single chimeric embryo that metaphysical questions arise. Outside of laboratory manipulations, it is difficult to know whether fusion of this type ever naturally occurs.

Part of the uncertainty regarding the mechanism of chimera formation stems from the rarity of the condition. Most cases of a single chimeric individual (rather than chimeric twins) are detected when infants present at birth with an intersex disorder and are determined to have a mixture of male and female cells. This excludes most cases of same-sex chimeras (approximately 50 percent of all chimeras)—which can go undetected for decades.[14] Yet not all cases of intersex disorders involve chimerism. For example, one in every one hundred thousand live-born individuals have both testicular and ovarian tissue, and of these, only approximately 10 percent involve chimerism[15] (with the remaining 90 percent being mosaics). In the most well-studied case of human chimerism,[16] the representation of cells with distinct genotypes varied considerably across different tissues, both making it difficult to exclude incorporation as a mechanism and (if fusion was the cause of the chimerism) making it difficult to predict when fusion was likely to have occurred.

UNIFYING CONCEPT 13: All singleton chimeric individuals could potentially be formed by incorporation of cells from a twin that does not survive, a mechanism that does not call into question the ontological status of the chimeric individual.

Philosophical Concerns Raised by Embryo Fusion

Formation of chimeras by embryo fusion is likely to be a rare event, yet due to the propensity of early embryonic cells to adhere to each other and to participate in a unified embryonic trajectory,[17] it remains a realistic possibility. Just as in the case of embryo splitting at the two-cell stage discussed above, if embryo fusion *ever* occurs, regardless of how rare this event may be, the philosophical issues remain. In light of our consideration of embryo splitting, what is the most reasonable view of cases (real or merely theoretical) where two living embryos fuse to generate a single chimeric embryo?

Based on experimental models of chimerism,[18] the embryonic trajectory initiated at sperm-egg fusion clearly persists; that is, the development of the fused embryo proceeds without perceptible delay from the point of fusion onward. This observation could suggest that, just as in the case of twinning, one of the original embryos survives the fusion event. However, continuing along the same developmental trajectory is not *sufficient* evidence to conclude to the persistence of a specific individual. For example, if a morula-stage embryo is entirely dissociated, it is clearly *dead*; the embryo has been destroyed, and what is left is a collection of cells derived from the embryo. However, if the dissociated cells are reaggregated, they are in proximate potency to being a living organism and are therefore capable of undergoing a substantial change to generate a living embryo, which resumes development where the original embryo left off, prior to dissociation; that is, the embryo formed by reaggregation continues development from the morula stage.[19] Yet despite containing the same cells that were present in the original embryo prior to dissociation and even though the newly constituted embryo proceeds along the same developmental trajectory, an embryo formed by reaggregation of cells would be a new (albeit very similar) individual, not a "reanimation" of the dead embryo.[20]

Chimeras formed by embryo fusion are similar to the case of embryos formed by aggregation of embryonic blastomeres, just described. Indeed, chimeric embryos are often produced in the laboratory in exactly this way; by dissociation (and therefore, destruction) of two embryos and aggregation of their cells to generate a single individual.[21] Yet whether embryos are first dissociated or not, following fusion, the genetic and cellular basis for the developmental trajectory of the resulting embryo (i.e., the developmental program) has been radically altered, compared to the development of the initial two embryos.[22] This is especially evident in the case of mixed-sex chimeras, where the initial development of each embryo along a sex-appropriate path is abandoned and the chimeric embryo matures as an individual with some degree of intersex characteristics.[23]

Importantly, the new developmental trajectory that is assumed following *fusion* of a male and female embryo differs in significant ways from the trajectory of intersex individuals arising as a consequence of cell *incorporation*, described above. Incorporation of cells from an embryo of the opposite sex results in reproductive malformations while the overall developmental trajectory remains unaltered. However, in the case of embryo fusion, the trajectory of the *entire* embryo is now governed by a new, chimeric developmental program, with novel developmental outcomes arising from the interaction of the two distinct genomes. For example, if one embryo produces high levels of a growth factor for bone but low levels of the receptor for that factor, with the opposite being true for the second embryo, neither would mature to be an exceptionally tall individual. However, if they fuse, the new, chimeric embryo is likely to have increased bone growth that results in an unusually tall individual at maturity, precisely due to the interaction of its two constituent genomes.

This suggests that upon fusion, the organizing principles governing the development of the two initial embryos are no longer operative, and that a new organizing principle—that of an intersex individual in cases of male-female embryo fusion—now governs the ongoing development of the chimeric embryo. As when the zygote is formed at fusion of sperm and egg, upon fusion of two independent embryos, a new entity with distinct molecular/cellular composition and distinct developmental behavior comes into existence.[24] Because the genetic and cellular control of development is radically altered *for the entity as a whole*, it is likely that upon fusion, both of the initial embryos die and their cells combine to form a new individual with a distinct developmental program and, therefore, a new human substantial form.

Although several of the philosophical concerns discussed above in the context of twinning also apply to chimeras generated by fusion, the response to these concerns is identical. Embryos that subsequently fuse are not virtual humans, but rather human individuals undergoing two independent trajectories of development, who die upon fusion. The fact that the two initial embryos do not leave corpses behind does not indicate the embryos somehow persist; the cells of the initial embryos have undergone a substantial change to generate a distinct, third individual. Finally, the parents of the original embryos, as the intentional actors in a procreative event, are the parents of the two deceased embryos as well as the third, chimeric embryo.

> UNIFYING CONCEPT 14: When two living embryos fuse to form a single chimeric individual, a substantial change occurs with the original embryos ceasing to be and a new individual coming into existence. However, the formation of an individual in this manner does not raise significant concerns regarding the ontological identity of the newly formed individual.

WHY SCIENTISTS ARE CONFUSED
ABOUT EMBRYOS

Many of the questions that have been raised by philosophers regarding human embryos can be addressed by a combination of accurate scientific observation and clear philosophical reasoning. Yet philosophers are not the only thinkers confused by the embryo. As we have seen with the coining of the term "preembryo" by a prominent biologist,[1] scientists can also draw inaccurate and misleading conclusions about embryos. Indeed, many of the most confusing statements regarding the beginning of human life and the nature of human embryos originate with scientists. For example, scientific authors frequently use the term "totipotent" in an inconsistent and misleading manner. These misapplications fall into four general classes: (1) equating *participation* in development with the ability to independently *generate* a developmental sequence, (2) equating the ability of groups of cells to *collectively* generate a developmental sequence with totipotency of individual cells, (3) equating the expression of early embryonic molecular markers with totipotency, and (4) taking a partial or superficial resemblance to an embryo as evidence for totipotency.[2] All of these inaccurate usages of the term create considerable confusion, both within the scientific community and within the greater public that relies on scientists to accurately interpret scientific results.

Scientific confusions about the beginning of human life are not limited to totipotency and "preembryos," but extend to the question of human value as well. The noted embryologist Scott Gilbert has stated categorically, "I don't know when personhood begins, but I can state with absolute certainty there is no consensus among scientists,"[3] despite the overwhelming consensus of the scientific literature on the issue of when life begins[4] and the logical implications of that consensus for human personhood (see below). In light of such misleading representations of the scientific evidence and such poorly reasoned conclusions from the facts, it is important to ask: Why are scientists so confused about the embryo? Moreover, why do scientists routinely deny human status to the embryo by conducting embryo-destructive research? Here, I will briefly consider the challenges presented by the culture of science and how this culture promotes unethical human embryo research.[5]

Biomedical Research and Ethics

The confusion of scientists about embryos exists as part of a much wider confusion within the scientific profession about ethics and human life. For much of the history of science, there were no formal ethical guidelines for human research, with research topics being largely selected based on the importance of the question or the curiosity of the researcher. In the wake of the research atrocities committed by the Nazi government during World War II, ten basic principles of ethical research practice were articulated as the Nuremberg Code.[6] The Nuremberg Code was subsequently formalized in the Declaration of Helsinki,[7] which became the basis of legal regulations governing medical research in the United States,[8] Europe,[9] and elsewhere.

The Helsinki principles stipulate that research involving human subjects (including embryonic and fetal human subjects) be directed toward the best interest of the research subject and conducted in a manner that protects the subject's "life, health, dignity, integrity, right to self-determination, privacy, and confidentiality."[10] The Helsinki Declaration does not make a principled argument regarding what it means to protect a patient's interests, but rather it takes the form of a list of rules that presupposes the correctness and universality of the basic underlying principles. Consequently, it represents a deontological approach to ethics; that is, certain ends are accepted as "good," and rules are established to accomplish those ends.

Yet in the current practice of biomedical research, the Helsinki principles are by no means universally accepted. Arguments are routinely made for research protocols that do not respect the life, health, and dignity of research subjects, due to the presumed value of such research for society. Examples include proposals to abandon the dead-donor rule in order to increase the supply of transplantable organs,[11] or experiments in which human embryos are destroyed to facilitate research into human infertility and human disease.[12]

The novel reinterpretations of historically accepted research norms noted above often are based on consequentialist arguments—that is, that the overall consequences of an action determine its moral value.[13] Starting with the claim that destructive experimentation on human embryos will advance human knowledge and benefit human patients, such research is believed to be justified. Yet how does such a consequentialist argument balance the interests of one segment of society against another when they are in conflict? Can a research program be justified if it disregards the dignity of some individuals for the benefit of others? And who will be empowered to make such decisions? In the absence of a consensus on what constitutes "ethical" research, a consequentialist perspective reduces to a purely numerical calculation; if the number harmed is less than the number benefited, the research is justified. Clearly, such a view defies the principles of justice, liberty, and equality and therefore does not provide a compelling or universal answer to the question of how to ground ethical research guidelines.[14]

The Scientific Method Is Inherently Utilitarian

Part of the appeal of consequentialist arguments for scientists is their structural similarity to the scientific method; various courses of action are evaluated based on a quantifiable numerical outcome (the greatest "benefit" to society). Yet the scientific method is clearly not an ethical framework for decision making. It is merely a precise way of making observations and drawing valid inference from them. The goal of a scientific investigation is to acquire information, and consequently, the primary measure of the "quality" of a scientific investigation is *utility*; that is, research is evaluated based on the quantity, accuracy, and applicability of the data it generates. Decisions to pursue a specific line of scientific research are often made on the basis of utilitarian considerations regarding the likelihood that the research will yield "useful" results that will subsequently enable successful competition for new research funding.

In addition to the tendency of scientists to make decisions based on utilitarian considerations, the scientific method *itself* is not neutral to the utility of scientific findings. Consider, for example, an experiment that has two potential outcomes, with outcome A being a "positive" finding (i.e., altering a specific parameter alters the phenomenon under investigation) and outcome B being a "negative" finding (i.e., nothing changes when you alter the experimental conditions). As noted above in the discussion of experimental tests of cell potency (see chapter 2, under "How Long Does Totipotency Persist?," especially n. 16), the negative finding B proves *nothing* and is therefore not useful in refining the experiment or explaining the phenomenon under investigation. In contrast, the positive finding A is useful because it both provides an explanation and suggests new experiments. Consequently, "useful," positive findings advance the scientific method, while negative findings with no utility do not, and in this sense, the scientific method *itself* has a strongly utilitarian dimension.

This is not to suggest that *scientists*, as human persons, are not obligated to make sound moral judgments about their research or that scientific findings cannot have moral or immoral applications. It simply illustrates that science, as science, is fundamentally utilitarian, providing no guidance in the domain of ethics beyond the precept that whatever has the greatest utility also has the greatest worth.

In modern times, the application of the scientific method to biomedical questions has been very effective.[15] The impact of biomedical investigations on human health, combined with the fact that the "quality" of the research has nothing to do with morality, has led to the view that ethics should not constrain biomedical research. This is particularly true in the area of destructive human embryo research, where the perceived value of the embryo is small and the perceived public benefit is large. Christopher Tollefsen characterizes proponents of this view as believing that "because some forms of inquiry promise such significant benefits to human beings, those pursuing the relevant research should not be hindered by 'ethics,' for the benefits would outweigh any possible negative consequences of the research."[16]

Within this context, it is important to appreciate that human embryos are both an exceptionally interesting scientific subject and an exceptionally useful tool. For example, production of embryos in the laboratory as a means of treating infertility has been a very lucrative technology, with a market size estimated at over $22 billion in 2015.[17] Similarly, while medical applications of human embryonic stem cells have lagged behind the applications of other sources of human stem cells,[18] the current global market for stem cell therapies is estimated to be $33 billion.[19] At the opposite end of the spectrum, the Guttmacher Institute estimates 926,200 abortions were performed in the United States in 2014, at an average cost of $470,[20] making the destruction of embryos and fetuses a multimillion dollar business. Clearly, there is ample economic pressure for physicians and scientists to view the embryo as nothing more than an attractive economic resource. Adding to the direct earning potential of human embryo destruction and research, human development is also a scientifically fascinating topic, with tremendous potential for making important scientific discoveries. In a profession where utility largely prevails over ethical considerations, it is not surprising that human embryos are not given a privileged moral status by scientists.

An important implication of the lucrative aspects of embryo-destructive procedures is that many of the scientists and physicians who seem most qualified to offer society a factual answer to the questions of when human life begins and what constitutes an embryo are subject to a significant conflict of interest. For example, in an opinion piece recently published in the Los Angeles Times, Dr. Richard Paulson, a physician specializing in infertility treatment, states categorically, "A preimplantation embryo has the theoretical potential to become a human life, but it cannot be considered on the same moral plane as a human life." Yet the author acknowledges that some answers to the question of when human life begins would have a profound impact on his own livelihood, stating,

"If a pre-implantation embryo were to be considered a human being, then . . . fertility clinics would not be able to function."[21]

In most contexts, the opinions of individuals with significant conflicts of interest are excluded from an unbiased consideration of a topic. Although embryologists (such as Dr. Gilbert, quoted above) and fertility specialists (such as Dr. Paulson) may be of the opinion that there is no consensus on when human life begins, scientists whose livelihood is not directly linked to human embryo experimentation quite frequently conclude the opposite, holding that human life clearly begins at sperm-egg fusion. Indeed, every major medical text on human embryology and dozens of peer-reviewed scientific papers openly acknowledge this fact,[22] often with direct statements such as, "Most readers of this review originated from a sperm-egg fusion event."[23]

> UNIFYING CONCEPT 15: The production, manipulation, and destruction of human embryos is a highly lucrative business. Consequently, the objectivity of scientists and physicians directly involved in embryo research can be significantly compromised on topics that potentially impact their financial and professional interests. Claims that there is no consensus on the moral status of the embryo or on when human life begins must be interpreted in light of such potential conflicts of interest.

Due to the complexity of biomedical research, society increasingly depends on the judgments of scientists and physicians to evaluate the moral character of research. In light of this dependence, it is important to understand the nature of biomedical research itself and the personalities selected for by the scientific profession, both to appreciate the challenges presented by the culture of science and to work effectively with the cultural elements that can potentially promote a more accurate view of the embryo.

The Nature of Biomedical Research and Researchers

Although basic biomedical research is focused almost exclusively on the morally neutral goal of acquiring knowledge, the means by which facts are acquired and the motivations of the scientist conducting the research have clear moral dimensions.[24] And basic biomedical scientists are driven by a complex and sometimes counterintuitive set of motives that can affect the moral character of the research, the researchers, and the institutions at which they work.[25] In considering the research culture, it is instructive to view the scientific enterprise through three distinct lenses: (1) science as a business, (2) science as a public service, and (3) science as art.

(1) In many ways, basic scientific research is simply a business: scientific results are the "products" that enable scientists to obtain research funding, which in turn supports the ongoing work of their laboratories and (in many cases) their own salaries. A successful business requires both a quality product and a "market" for that product, and the scientific enterprise is no exception. Importantly, the primary market for scientific results is not the greater society that funds the research, but rather other scientists who, through the mechanism of peer review, reward those studies they judge to be of high quality with continued grant funding. The general public has almost no influence on the success (or failure) of a scientific enterprise.

A consequence of this closed system of peer review is that the utilitarian perspective inherent in the scientific method is strongly reinforced. Studies with the greatest explanatory power are the most marketable studies, precisely because they are perceived as being of high quality by other scientists. Yet surprisingly, the competitive nature of research funding also strongly reinforces the virtue of fortitude, or "firmness in the face of difficulties and constancy in the pursuit of a good end."[26] Typically, scientists are a stalwart group, who are willing to work long hours and endure considerable frustration in the hope of obtaining high-quality results.

A second, and perhaps more insidious, effect of the business aspect of science is how reliance on research funding can erode the university as an institution of higher learning. Scientific research often constitutes a significant financial resource for universities. For example, based on the annual financial report prepared by my own institution, grant income in 2016 was in excess of $367 million,[27] a relatively modest figure that ranks the university fortieth out of more than two thousand five hundred institutions receiving funding from the National Institutes of Health that year.[28] However, even this modest grant portfolio was $50 million larger than the total income derived either from tuition or from state appropriations to the university, as given in the same report.

Modern research universities have increasingly come to rely on grant funding as a source of revenue, and this reliance can profoundly affect academic freedom and the pursuit of interdisciplinary research. When the "success" of faculty is measured almost exclusively by the utilitarian criterion of grant dollars,[29] researchers are strongly incentivized to pursue only lucrative projects and to avoid any activities that do not directly result in increased grant funding—including teaching, service, collaboration, public education, or the search for meaning and truth. This serves to both isolate scientific researchers from their colleagues in the humanities and to elevate the perceived importance of scientific research as the only form of knowledge that matters for the financial health of the institution.

(2) Basic scientific research can also be seen as a public service, similar to education or medicine. And like other public servants, scientists are strongly motivated to make positive contributions to society. Even the most successful research programs only rarely advance human welfare directly, yet most scientists remain firmly convinced that their research will ultimately benefit humanity. This conviction demonstrates a sincere dedication to the virtue of justice, or "the will to give to others what it is rightly their due."[30] A recent comparison of altruistic versus economic motivations for research across different nations consistently ranks the United States among the most altruistic nations across a broad spectrum of research activities.[31]

(3) Finally, and perhaps somewhat surprisingly, basic scientific research can also be viewed as an artistic endeavor. Many scientists passionately love their areas of research and invest considerable effort in designing studies that are elegant and beautiful. For example, scientists routinely produce illustrations of their work that are both informative and aesthetically pleasing (figure 7.1). Even in the absence of such stunning illustrations, many arcane studies without any immediate practical application are nonetheless highly admired among scientists and upheld as both important and "elegant."[32] For scientists, as for all humans, love of beauty is a powerful path to the true and the

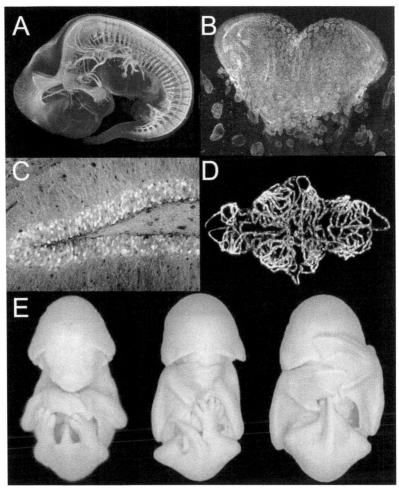

Figure 7.1. Embryology as art. (**A**) The peripheral nervous system of a mouse embryo. (**B**) The eye of a larval fruit fly. (**C**) Cells within the hippocampus of a genetically engineered mouse embryo. (**D**) The vasculature of the brain in a zebra fish. (**E**) Three bat embryos that are part of a scientific staging series. Images were obtained from Nikon's Small World competition (https://www.nikonsmallworld.com/) and are used with permission.

good, and the aesthetic aspects of science can serve to lead even the most jaded scientist to a broader appreciation both of the natural world and of human value.

In my experience, despite the virtues reinforced by the culture and practice of science, the scientific profession tends to attract individuals who are not strongly inclined to think about moral issues and are therefore largely unsuited to make sound moral judgments.[33] Scientists who are unconcerned with how the public views their research understandably make little effort to reflect upon the broader impact of their research or the ethical issues it raises. This lack of concern is reinforced by the closed system of peer review, which incentivizes scientists to strive for "the best" science, a goal that does not entail a meaningful consideration of ethics.

> UNIFYING CONCEPT 16: The utilitarian nature of science and the incentives provided by a competitive funding environment within a closed system of evaluation strongly motivate scientists to regard the human embryo as nothing more than a useful experimental tool. However, the virtues of fortitude and justice are also strongly reinforced by the scientific culture and can be a useful starting point for promoting a meaningful dialogue about the moral worth of the embryo.

Moving beyond a Utilitarian Approach

As for all human endeavors, the goals of biomedical research cannot be defined independently of the best interests of either society or the human person conducting the research. In light of the lack of consensus on what constitutes ethical research, how are research guidelines to be established? Addressing this question requires a clear view of what virtues will promote the ethical practice of science and a clear strategy for encouraging these virtues within the scientific community.

Virtues have been historically defined as admirable habits, or those that provide evidence for moral excellence. The concept that, for any given class of action, virtuous behavior represents the mean between two extremes was originally put forward by Aristotle in the *Nichomachean*

Ethics and is a common basis for judging specific behaviors. For example, many would agree that courage lies between cowardice and rashness; modesty lies between false humility and pride; a healthy diet lies between indulgence and starvation. The Catholic Church has applied similar reasoning in defining the four cardinal virtues that guide moral behavior (prudence, justice, fortitude, temperance).[34]

As noted earlier, the fact that biomedical research requires perseverance in the face of adversity strongly promotes the virtue of fortitude. Similarly, the widely held belief among scientists that publicly funded research should benefit the public reinforces the virtue of justice. Because success in the scientific profession is strongly correlated with both of these virtues, they are well represented among scientists. In contrast, many of the ethical failings in biomedical research reflect the fact that the profession does little to promote (and, indeed, is actively antagonistic to) the two remaining cardinal virtues: temperance and prudence.

Temperance is the virtue that "ensures the will's mastery over instincts and keeps desires within the limits of what is honorable."[35] The virtue of temperance is most strongly associated with the control of bodily appetites, but it applies to intellectual appetites as well. And in modern times, many of the most ethically problematic research approaches reflect a failure to adequately constrain intellectual curiosity.

Scientists tend to reflect only trivially on the question of whether something should be done, seeking instead to push the boundaries of what is technically possible as a means of satisfying their intellectual curiosity. For example, the lead author of a sensational study producing mice with two genetic fathers justified the research saying, "It has been a weird project, but we wanted to see if it could be done."[36] And satisfaction of intellectual curiosity is the motivation behind a large number of experiments—both ethical and unethical.

The problems arising from an unconstrained pursuit of curiosity are exacerbated by the almost complete absence of prudence in the scientific community. Prudence is defined as "the virtue that disposes practical reason to discern our true good in every circumstance and to choose the right means of achieving it."[37] Scientists, like most people, do not generally choose to pursue a course of action they know to be immoral. Rather, they simply never ask the question of whether their research is accomplished by moral means because the question of prudence is not considered a meaningful question.

Without ethical guidance from the profession and without a strong inclination to either temperance or prudence, research topics and research methodologies are selected largely on utilitarian grounds. Human cloning, human stem cell research, human genome editing, and a wide range of other morally questionable research areas reflect little more than the interest of the questions and the power of the techniques, independent of any serious ethical consideration. This is not to suggest that scientists do not have charitable motivations, but only to note that without a well-articulated ethical framework, individual scientists are left to make ethical judgments entirely on their own. Thus, if an individual scientist does not believe an embryo is a human person and if, in their professional judgment, research on human embryos will benefit humanity, this is sufficient justification to conduct the research.

> UNIFYING CONCEPT 17: The virtues of temperance and prudence must
> be inculcated into the culture of science to restrain unethical research, to
> promote a wider appreciation of the moral worth of human embryos, and
> to direct appropriate policy regarding human embryo experimentation.

Practical Suggestions

It is important to ask what we, as members of society, can do to help promote the virtues of temperance and prudence within the scientific profession. Three practical suggestions have been offered.[38]

1. **Science literacy:** Biomedical research impacts society in important ways, and as citizens, we must accept *personal responsibility* for promoting basic scientific literacy, so that we can intelligently evaluate biomedical research and participate in the formulation of science policy that is based on sound moral judgments.
2. **Community-based scientific policy:** Rather than allowing science policy to be established by a handful of "experts" and government administrators, we should urge that policy be actively debated by the

constituents who are most directly affected and by the greater community. This could in part be accomplished by encouraging individual scientists to actively participate in professional societies, thereby promoting a broader representation of views. Efforts to involve community members (patient groups, religious groups, and ethnic groups[39] impacted by biomedical research) in decision making also serves to bring ethical and moral considerations held by the greater public into the discussion of research policy.[40]

3. **Educating the next generation of scientists:** The workforce employed in the fields of science, technology, engineering, and math (STEM) continues to expand, with an estimated 20 percent of all US jobs requiring a high-level STEM education.[41] Consequently, there are ample opportunities to better incorporate greater ethical education into STEM curricula. The role of biomedical research in society and the role of ethics in formulating a sound research design should be actively addressed, beginning at the high school level. For those directly entering the biomedical research field, the National Science Foundation offers funding for ethics education.[42] Proposals should be designed to specifically articulate how the virtues of prudence and temperance can help shape biomedical research in the context of a just society to promote human flourishing.

An important (albeit somewhat uncomfortable) implication of these practical suggestions is that nonscientists play a critical role in constraining and directing biomedical research. To enact this role, citizens must actively participate in the establishment of science policy and insist that future scientists receive adequate education in ethics. To serve these goals, citizens must be competent to evaluate the ethics underlying biomedical technologies, without relying critically on scientific experts who may be biased by financial or professional interests.

UNIFYING CONCEPT 18: The public can play an important role in regulating human embryo experimentation and educating future generations of scientists on key ethical issues. Yet these roles rely critically on citizens assuming *personal responsibility* for achieving greater scientific literacy on topics related to the human embryo.

CHAPTER 8

THE EMBRYO IN A LARGER CONTEXT

Embryos, even when they consist of only a single cell, are not like other human cells. At all stages of life, human beings are defined by the capacity for body-wide integration.[1] Important features of organismal integration are autonomy and self-integration; that is, the bodily functions of human beings are controlled and orchestrated by the human being itself to maintain its health and survival. Autonomous self-integration for the sake of the body as a whole is a critical diagnostic feature of both the beginning[2] and the end[3] of human organismal life. As noted elsewhere,

> Our understanding of when human life both begins and ends depends critically on the distinction between a living human being and living human cells. Human beings are multicellular organisms that autonomously integrate the biological activities required for continued health and survival of the organism as a whole. While aggregates of cells can have complex properties, they do not show such global integration of function.[4]

But how are we to view the "whole" toward which the structures and functions of the embryo are ordered? And what ultimately establishes a *human* ordering within an embryo, when the molecules that constitute an embryo could just as easily be part of a tumor or a chimpanzee?

The bias of modern science is to dismiss the metaphysical question of what an embryo is as irrelevant, and instead focus on the practical questions of how an embryo is constructed (efficient cause), what it is made of (material cause), and how it can be manipulated (a purely utilitarian consideration). Yet as the history of basic research amply illustrates, proceeding without sound ethical guidance is a path fraught with danger. And such guidance requires an answer both to the metaphysical question of what the embryo actually is and to the ethical question of what obligations we owe the embryo.

An Aristotelian View of the Embryo

Aristotle addresses the question of what constitutes a whole human being by appealing to the theory of hylomorphism and the concept of substantial form. In a hylomorphic view, substantial form is the cause of the embryo and the reason it is distinct from a tumor or a chimpanzee.[5] Substantial form remains constant throughout human life, despite the complex and ongoing changes in the structure and composition of the body over developmental time.[6] But what is substantial form, in concrete terms? In precise philosophical language, substantial form is "the principle of act in relation to prime matter that makes something be what it is, most fundamentally."[7] Yet as modern thinkers, we struggle with such an abstract definition. To address this difficulty, it is instructive to consider a more familiar, scientific perspective on substantial form:

> From a biological perspective, the simplest way to consider substantial form is to ask: What, in concrete terms, makes a living human different from a simple pile of organic molecules that is identical to those found in the body?
>
> What differs between a human and a pile of chemicals is not what they are made of but how that material is organized or "formed," and substantial form is the cause of the observed organization. The precise relationships and interactions between our molecular parts do not arise from the parts themselves, otherwise there would be no difference between a human and a pile of molecules identical in composition to a human being.

Rather, the matter comprising a human is ordered by a specific set of rules that is distinct from the rules ordering those same molecules in a non-living pile. The rules governing molecules in a human are not a physical thing in themselves; i.e., some "Ingredient X" that is added, like pixie dust, to animate the body. This view is called "vitalism" and is in full display in Mary Shelley's Frankenstein, where the addition of the mysterious electrical force brings the monster to life. Nevertheless, the rules must be real, and really present in living things, else the unique order of life has no explanation.

A developmental biologist sees these rules as practical matters of ligand-receptor binding affinities, concentration gradients, intracellular diffusion constants and other principles that govern how molecules function within cells and how cells communicate during formation of the mature body. Yet collectively, the rules or principles governing the life, function and development of a particular living being constitute the substantial form of that individual.[8] To emphasize that there is nothing mystical about substantial form, for most living things, if it were technically possible for an experimenter to start with a collection of inert organic compounds and establish all of the interrelationships between the molecules, cells, tissues and structures that are observed in a living body, this would "create" a soul that would subsequently be the cause of all the ongoing properties of life observed in the entity thus constituted. . . . Substantial form, in combination with the particular character or potency of properly disposed matter (the unique genome, transcriptome and proteome that contributed to a particular individual), constitutes a particular substance that we recognize as [an individual].[9]

Both Aristotle and Aquinas would argue that the *human* substantial form represents a special case,[10] yet the case for animal substantial form discussed above serves to illustrate the meaning of this term in language that is more familiar to the modern mind. Importantly, substantial form is not a mystical element added to ordinary materials such as carbon and phosphorous to make them human. It is an organizing principle that "enforms"[11] the matter constituting a human being at all stages of human life. And the integrated function of the embryo from the instant of sperm-egg fusion onward provides clear evidence for the presence of a human organizing principle. Therefore, when the product of sperm-egg fusion comes into existence with the capacity to undergo development, a human organism (i.e., human being) with a human substantial form also comes into existence.

The difference between a zygote and a complete hydatidiform mole (CHM) discussed above serves to illustrate the centrality of substantial form in determining the nature of a living entity. The acts of both the zygote and the CHM are caused by their underlying substantial forms, and these forms are fundamentally different. A zygote possesses a *human* substantial form that manifests through the process of development. A CHM possesses a *cellular* substantial form that manifests through formation of a tumor—that is, an aggregate of human cells. While death can cause a substantial change that converts a living human being into an aggregate of living human cells, this change is manifested by a loss of organismal function; in death, the cells of the body no longer function as an integrated whole. In contrast, a CHM never exhibits a globally integrated pattern of behavior that rises above the level of a cell (i.e., it never exhibits the higher level of integration seen in either a tissue or an organism) and therefore does not possess a human substantial form.

UNIFYING CONCEPT 19: The organismal nature of the human embryo, beginning at the instant of sperm-egg fusion, is caused by the human substantial form. Substantial form is not a "mystical" quality, but rather the organizing principle that determines the specific nature or character of a human individual.

Ethical and Legal Status of "Preembryos"

The Aristotelian view that a human substantial form causes all the acts of the embryo and exists from the very beginning of life is by no means universally accepted. While some view embryos as immature human beings,[12] others consider them to be merely "potential" human beings, worthy of consideration but not of full human rights.[13] Importantly, the debate over the status of the human embryo is not merely academic. The advent of modern assisted reproductive technologies has brought human embryos to the attention of the legal system. Excess, frozen embryos produced by fertility clinics have frequently become the subject of litigation between the parents who created them. In some cases, courts have viewed such embryos primarily as property and judged that destroying them can be warranted to prevent forced procreation.[14] In other cases, courts have either viewed embryos as human beings[15] or concluded precisely the opposite.[16] In these situations, the term "preembryo" is often invoked to distinguish the limited consideration due to early human embryos, compared to later embryonic and fetal stages. For example, in 2005, Judge Kessler stated in *Jeter v. Arizona*, "Current scientific knowledge concerning embryonic development underscores the difference between a viable fetus in vivo and an eight-cell, three-day-old pre-embryo in vitro."[17]

The term "preembryos" and the distinction it purports to make have clearly not been adopted by the scientific community. A query of PubMed, the world's largest database of scientific publications, returns only 395 papers using the term "preembryo" over the last four decades, with many of these works being ethical commentaries rather than scientific research.[18] In contrast, the standard scientific terms used to describe the period of development Grobstein rechristened "preembryonic" (i.e., zygote, preimplantation embryo, morula, blastocyst) were used 36,435 times over the same period.[19] Yet even though "preembryo" is not an accepted scientific concept, the term is widely employed outside of the scientific profession. A recent search of the LexisNexis database of legal and business publications returned over 1,200 citations using the term "preembryo," including more than 200 from the last five years alone.[20] Given the range of opinions regarding the ethical and legal status of early human embryos and the significance of the issues that turn on these opinions, it is important to revisit Clifford Grobstein's argument in favor of the term "preembryo" in light of the scientific evidence and scientific reasoning presented here.

As noted in the introduction, in an article entitled "Biological Characteristics of the Preembryo," published in 1988,[21] Grobstein made a number of arguments in support of the claim that human embryos do not exist until formation of the primitive streak (approximately fourteen days following sperm-egg fusion). Grobstein's strongest argument is based on "individuation." He holds that because "preembryos" can be both split and fused (i.e., because monozygotic twinning and chimera formation are possible), early embryos lack "developmental individuality" and consequently "would not develop into an integrated and single adult."[22]

Yet this is patently false. It is a simple matter of fact that the vast majority of embryos that survive to live birth (greater than 99.6 percent) do precisely what Grobstein claims they would not do: develop into an integrated and single individual.[23] Moreover, this distinction reflects an inadequate understanding of potency. For example, experiments on animal cloning (first published in 1958[24]) clearly indicate that a portion of an adult individual can be split off and used to produce a second, genetically identical individual. This result demonstrates that parts of a mature human exhibit a remote, passive potency to become a complete human. Since it would have been theoretically possible to clone Dr. Grobstein at the time he coined the term "preembryo," his own reasoning would require that he had not yet achieved "developmental individuality" and was still in the "preembryonic" stage, but just didn't realize it. Finally, as we have seen, neither twinning nor formation of chimeras by embryo fusion calls into question the ontological identity of the embryo. Upon embryo splitting, the remote, active potency of early blastomeres to become a full and complete individual becomes proximate, and a substantial change occurs to generate a new individual. Upon fusion, the initial two embryos die, yet their surviving cells remain in proximate potency to undergo a substantial change and form a new individual.

Grobstein's second argument is that the "preembryo" must produce the first two cell types (trophectoderm and inner cell mass; TE and ICM) before it can be considered an embryo. Yet this assertion fails to acknowledge that all of the structures, molecules, and molecular interactions required to produce the first two cell types existed in the embryo from the beginning. Consequently, formation of TE and ICM provides evidence an embryo exists, but does not *produce* an embryo. Moreover, it is not clear from Grobstein's reasoning why formation of TE and ICM carries "embryo-forming" significance. If the mere production of new cell types were to signify that a new kind of entity has arisen, humans would undergo multiple changes as significant as the one from "preembryo" to embryo over the course of life—for example, at puberty, when mature sperm and egg cells are first produced. Yet we clearly recognize that adolescent and adult humans are not ontologically different kinds of beings but rather are different developmental stages of the same kind of being—namely, a human being.

Finally, Grobstein argues that because development is a continuum, "the advent of developmental individuality is only a step—albeit a major one—on the path to full individuality, with functional, behavioral, psychic, and social aspects still to be achieved."[25] Yet how or why this observation alters the nature of the embryo is unclear. At all stages of life, maturation is a continuous process. All of us were undoubtedly "pre-adults" before we achieved "behavioral, psychic, and social" maturity. Yet during childhood and adolescence, we did not have a significantly different nature or restricted value relative to the adults we ultimately became.

From these three arguments ("individuation," sequential formation of different cell types, and the continuous nature of maturation), Grobstein concludes, "The scientific account that has been given clearly characterizes the preembryo as qualitatively different in kind from the embryo into which it subsequently transforms." Yet this "scientific account" does not accurately represent the scientific facts (even those known at the time) and does not give a coherent account of early human development. Grobstein provides no evidence for a "qualitative difference in kind" that distinguishes early and later stages of human development and therefore does not justify the coinage of a name that explicitly sets "preembryos" apart from humans at more mature stages.

UNIFYING CONCEPT 20: Arguments made in support of the concept of a "preembryo" are logically inconsistent and scientifically unfounded, providing no evidence for a qualitative difference in kind that distinguishes early stages of human development from later stages.

Aristotle and Abortion

To support the conclusion that human embryos are human beings with inherent value, some authors appeal to the opinions of ancient thinkers on the value of prenatal life, in particular the views of Aristotle. Matthew Lu has considered Aristotle's views on abortion and infanticide in detail. As noted by Lu[26] and others,[27] Aristotle holds in *Politics* 7.1335b19–26 that abortion is permissible only "before the onset of sensation and life." In light of the modern biological evidence that the human embryo is a living human organism with a human substantial form from the point of sperm-egg fusion onward, Aristotle's statement has led Lu, among other commentators, to conclude, "Faced with the results of modern embryology, I think Aristotle would be compelled by his own (implicit) principle to extend the unlawfulness abortion [*sic*] to the moment of conception."[28]

Yet, upon reflection, Lu concludes that Aristotle's position on the embryo and the permissibility of abortion cannot be interpreted so simplistically.[29] Lu notes that both Aristotle and Aquinas held that animals exhibited a succession of souls over the course of development, with an initial "nutritive" soul being replaced by a "sensitive" soul and finally a rational, human soul. In Aristotle's view, only upon acquisition of the rational, human soul after forty days of development (in males) could an embryo properly be considered a human being. An alternative view of Aristotle's position is that the developing embryo possesses only a single human/rational soul from the very beginning, with the sensitive and rational powers existing as intrinsic powers of that soul that are actualized over time.[30] Yet in either view, Aristotle's statement in *Politics* 7 is consistent with the view that abortion is permissible while the embryo exhibits exclusively nutritive functions, before it acquires sensitive and rational functions.

How are we to interpret Aristotle's prohibition on abortion after acquisition of the sensitive faculties? Lu maintains that Aristotle's view of the impermissibility of abortion must be placed in the proper context, which includes his views on the permissibility of infanticide and the necessity of population control. Lu notes:

> We must recognize that Aristotle's entire discussion of abortion is taking place in a world where exposure/infanticide was widely practiced and that he seems to more or less take this for granted. This alone should lead us to realize that he is in no position to generate a strong prohibition against abortion.[31]

Given that Aristotle did not oppose infanticide and believed family size should be limited by law, even in cases where this required abortion, Lu argues[32] that Aristotle concludes abortion is warranted as a form of population control in communities where infanticide is not a permissible alternative. Indeed, it does not seem that any other conclusion can be drawn because, as Lu notes, "if one did not think it was wrong to practice infanticide on a newborn infant, what reason would there be to reject abortion?"[33]

Lu is warranted in urging a better appreciation of "just how distant our moral universe really is from Aristotle's," an appreciation that would "require us to do the hard work of rethinking large parts of normative structure of his thought in light of his deeper metaphysical commitments."[34] Aristotle existed in a world that accepted infanticide yet drew a distinction between a newborn infant or fetus (in Greek, a *brephos*) and a child that had been formally accepted into the family by the father in a ceremony known as the *amphidromia* (such an accepted child is a *pais* in Greek). Greek law considered killing of a formally "recognized" *pais* to be a crime, while allowing infanticide and abortion of an "unrecognized" *brephos*.

Although Aristotle's views on abortion and infanticide were clearly conditioned by his culture and by his limited understanding of human embryology, it is interesting to consider how he would have viewed the embryo had he possessed an accurate understanding of human development. Lu suspects that in light of modern embryology,

> Aristotle would be content to give up his view of embryological development as consisting of a succession of different types of souls. Rather, he would likely conclude that with nuclear fusion of the parental gametes the zygote becomes an ontologically distinct biological human individual, with latent sensitive and rational potentialities.[35]

Kevin Flannery has also addressed Aristotle's reasoning on this matter in some detail,[36] and Flannery and I have extended his analysis to include an accurate account of early human embryology,[37] with both analyses concluding that Aristotle would have viewed sperm-egg fusion, not nuclear fusion, as the onset of a distinct human substance with latent sensitive and rational potentialities. Yet in either view, the question remains: Is the conclusion that infanticide and abortion are licit consistent with the unique value Aristotle assigns to the rational human substantial form?[38]

Aristotle appeared to hold, as many do today,[39] that the relevant parameter governing the question of abortion/infanticide is the social/legal consensus on the moral value of the embryo. Killing a child (*pais*) is illicit due to the social recognition conferred by the father in the *amphidromia*, while killing a newborn (*brephos*), who has not yet benefited from such recognition, is licit. In both cases, the legitimacy of the action is determined by social norms, not by the ontological nature of the child itself.

This argument is analogous to the contemporary distinction between a human being (i.e., a living human organism) who can be legally aborted by the mother and a human *person* (i.e., a human being who is "recognized" as being the subject of human rights). Yet this social/legal distinction does not alter the fact that both *pais* and *brephos* are clearly human beings. Although Aquinas does not use the term "person" with the same legal connotation it carries in the modern world, he accurately notes that while each individual human being is a representative of a particular species, they are also individual human *persons* with a rational nature that confers upon them a unique value:

> So a special name is given among all other substances to individual beings having a rational nature, and this name is "person." Thus in this definition of person, the term 'individual substance' is used to refer to a singular being in the category of substance; "rational nature" is added to mean the singular being among rational substances. (*Summa Theologica* I, q. 29, a. 1, sed contra)

Thus, while Aristotle's conclusions regarding infanticide/abortion are consistent with the laws and normative values of his culture, this observation leaves open the question, taken up below, of how human value is best determined in light of the scientific and philosophical nature of the embryo.

Human Embryos and Human Value

The scientific evidence clearly indicates that a one-cell human organism, which is qualitatively distinct from the cells giving rise to it, forms immediately upon the fusion of sperm and egg. From both a scientific and an Aristotelian perspective, this single cell is a complete and living human organism—that is, a member of the human species. Yet despite the clarity of the scientific evidence, many find this conclusion difficult to accept. It can seem absurd to call a single cell a "human being," and consequently, many simply reject the evidence as irrelevant, and instead ask different questions: When does human life become *valuable*? When does human *personhood* begin?

Like the question of when life begins, the question of when human beings have moral status and a right to life has also been answered in many ways. The three most common approaches are to confer rights based on (1) some aspect of form and/or function (or ability), (2) social convention (or fiat), and (3) status as a human being (or nature).[40]

Historically in Western society, human "personhood" (or the possession of human value and human rights) accrues to human beings by virtue of the fact that they are human; that is, all humans are created equal and endowed with unalienable rights. Logically, this requires that since human beings from the one-cell stage onward have a human substantial form, all human beings also have identical human value. There have, of course, been significant and egregious departures from this universal rule, most notably in the cases of slavery, women's suffrage, and the unjust treatment of individuals with mental disabilities (including forced sterilization and involuntary experimental brain surgery). However, the principle of equal treatment of all humans has steadily expanded, such that in the modern Western world, slavery is not practiced, and women are free to exercise their rights as voting citizens. Notably, over 170 countries have ratified a United Nations "Convention on the Rights of Persons with Disabilities," which asserts that individuals with mental disability are owed the same respect and rights as all citizens.[41]

In contrast, some assert that an embryo is not a "person" because it is immature (based on age), or because it doesn't look like a human (based on appearance), or because it is not capable of specific human functions, such as "viability" or "consciousness" (based on ability).[42] However, the novel proposal that human beings at some stages of life are not "persons" and that human rights are legitimately based on age, appearance, or ability significantly alters what the term "human rights" means for all human beings. Importantly, this shift from the historic norm of Western society makes "human rights" merely an arbitrary matter of consensus, thereby changing a "right" to a "convention" that is defined by those in power, according to their will.

The principles of justice, liberty, and equality form the basis of law in all Western societies, and denying rights to the embryo based on ability or social consensus defies these foundational values. Most of us reject linking rights to either ability or consensus as repugnant. It defies our basic sense of justice to envision a world where the strong, the beautiful, and the intelligent have a right to life, liberty, and the pursuit of happiness while the weak, the plain, and the slow do not. Similarly, most of us find repulsive the idea that a simple plurality of opinion can decide, as it did in Nazi Germany, who has rights and who does not.

The principles of justice, liberty, and equality impose clear duties on society with respect to human rights. Justice requires that humans receive what they are rightly due and rests on the foundation that *all* humans are the subject of natural rights. Liberty requires independence from government or private interference and rests on the foundation that the freedom (or life) of individuals cannot be restricted without due cause. Equality requires that all people be subject to identical treatment under the law and rests on the foundation that, regardless of age, ability, or social status, all humans have *the same* human rights. The only way of viewing human rights that does not offend the principles of justice, liberty, and equality is that rights are *unalienable for all human beings*, in other words, that we have rights only and always because we are humans. Such unalienable rights apply equally to humans at all stages of maturity, including the zygote stage.

The fact that human rights adhere to human beings by virtue of what they are has clear consequences for our duties to the embryo. As individuals and as a society, we have an obligation to respect and defend the life and dignity of the embryo. Embryos must not be bought or sold; they are human individuals, not property. Embryos must not be used as a means to an end; they are ends in and of themselves. Embryos must not be harmed or destroyed to achieve any purpose, no matter how noble that purpose may be;[43] they are persons, not tools. Finally, embryos must not be discarded or abandoned because they are "unwanted" or "unnecessary"; they have inherent dignity and worth that is equal to that of any other human being.

UNIFYING CONCEPT 21: The only basis for assigning moral status to human beings that does not defy our foundational values of justice, liberty, and equality is to assign worth based on the intrinsic nature of all humans as *human beings*. The scientific facts of human development indicate that the life of an individual human with a human substantial form begins at sperm-egg fusion. In identical twinning, this individual embryo splits to give rise to a second embryo, who is also (from the instant of splitting) a complete, albeit immature, human being with intrinsic value—that is, a human person.

CONCLUSION

Multidisciplinary Unifying Concepts
for Considering the Human Embryo

This analysis of human embryology and human twinning identifies twenty-one unifying concepts—drawn from the disciplines of biology, philosophy, sociology, and theology—that provide a three-dimensional context for interpreting the nature and value of the human embryo.

1. Based on objective, scientific criteria, the life of an individual human being unambiguously begins at sperm-egg fusion.
2. The moral and ethical consideration due to a human embryo is based on the underlying nature of the embryo itself, rather than on the mature state it will ultimately achieve.
3. Totipotency is a property of a single cell, and a mature human oocyte is a critical component of totipotency. Therefore, only a cell directly derived from an oocyte (i.e., the zygote) or the immediate progeny of the zygote (if separated from the embryo as a whole) can be a totipotent human embryo.
4. Human gametes are highly specialized cells that are predisposed to form a human being upon fusion, yet not all cells produced by the ovaries and testes are functional gametes.

5. The capacity for development is the defining (i.e., essential) feature of a zygote, and formation of the first two committed cell types in a temporally and spatially integrated manner is the minimum criterion for identifying that this capacity exists.

6. If the necessary structures (molecules, genes, etc.) required for production of the first two committed cell types do not exist in an entity from the beginning, the entity is intrinsically incapable of undergoing development as an organism and is therefore not a human being.

7. There is clear scientific evidence that the one-cell embryo or zygote initiates a developmental trajectory; that is, the zygote is manifestly a human organism. Therefore, twinning at the two-cell stage or later does not call into question the ontological status of the original embryo as a complete and individual human being.

8. Proximate active potential to develop to mature stages of human life is the criterion for being a human organism—that is, a human being at the beginning of the human life span.

9. Monozygotic twinning most closely resembles regeneration or wound healing of an injured human organism; development proceeds without pause along the trajectory initially established by the zygote, indicating that one of the two twins produced by splitting is ontologically identical to the original zygote.

10. Sexual intercourse is intrinsically (or naturally) ordered to reproduction and therefore parenting of a child can only be an intentional act. In contrast, monozygotic twinning is a random disruption of actions that are naturally ordered toward development, not reproduction. Therefore, in cases of monozygotic twinning, the parents of the original embryo are the parents of the twins.

11. Monozygotic twins are unlikely to be truly "identical." Distinguishing them from each other and determining which twin is ontologically identical to the original zygote is an epistemological question. Importantly, human individuals are not made to be distinct by distinguishing differences between them.

12. Similar to mature individuals who receive a blood transfusion or an organ donation, chimeric individuals who are produced by incorporation of cells from a twin (whether the twin survives or dies) remain ontologically the same individual before and after incorporation.

The incorporated cells are governed by the unifying principle (or substantial form) of the individual in whom they reside.

13. All singleton chimeric individuals could potentially be formed by incorporation of cells from a twin that does not survive, a mechanism that does not call into question the ontological status of the chimeric individual.

14. When two living embryos fuse to form a single chimeric individual, a substantial change occurs with the original embryos ceasing to be and a new individual coming into existence. However, the formation of an individual in this manner does not raise significant concerns regarding the ontological identity of the newly formed individual.

15. The production, manipulation, and destruction of human embryos is a highly lucrative business. Consequently, the objectivity of scientists and physicians directly involved in embryo research can be significantly compromised on topics that potentially impact their financial and professional interests. Claims that there is no consensus on the moral status of the embryo or on when human life begins must be interpreted in light of such potential conflicts of interest.

16. The utilitarian nature of science and the incentives provided by a competitive funding environment within a closed system of evaluation strongly motivate scientists to regard the human embryo as nothing more than a useful experimental tool. However, the virtues of fortitude and justice are also strongly reinforced by the scientific culture and can be a useful starting point for promoting a meaningful dialogue about the moral worth of the embryo.

17. The virtues of temperance and prudence must be inculcated into the culture of science to restrain unethical research, to promote a wider appreciation of the moral worth of human embryos, and to direct appropriate policy regarding human embryo experimentation.

18. The public can play an important role in regulating human embryo experimentation and educating future generations of scientists on key ethical issues. Yet these roles rely critically on citizens assuming personal responsibility for achieving greater scientific literacy on topics related to the human embryo.

19. The organismal nature of the human embryo, beginning at the instant of sperm-egg fusion, is caused by the human substantial form.

Substantial form is not a "mystical" quality, but rather the organizing principle that determines the specific nature or character of a human individual.

20. Arguments made in support of the concept of a "preembryo" are logically inconsistent and scientifically unfounded, providing no evidence for a qualitative difference in kind that distinguishes early stages of human development from later stages.

21. The only basis for assigning moral status to human beings that does not defy our foundational values of justice, liberty, and equality is to assign worth based on the intrinsic nature of all humans as human beings. The scientific facts of human development indicate that the life of an individual human with a human substantial form begins at sperm-egg fusion. In identical twinning, this individual embryo splits to give rise to a second embryo, who is also (from the instant of splitting) a complete, albeit immature, human being with intrinsic value—that is, a human person.

Vital Questions regarding Human Life at the Embryonic Stage of Development

These concepts establish a view of the embryo that is consistent with the scientific facts yet also addresses the three vital questions raised at the outset of this investigation from a wider base of human understanding.

When does human life begin? The question of when human life begins is largely a matter of science but is also addressed by the Aristotelian concept of substantial form: a human being is caused by the presence of a human substantial form that comes into existence at the instant of sperm-egg fusion. Clear evidence for the presence of a human organizing principle is given by the organismal events the embryo autonomously initiates immediately after sperm-egg fusion, with the formation of the first two committed cell types in an appropriate developmental sequence being necessary and sufficient evidence to distinguish an embryo from a nonembryo.

Is a human embryo a human individual? The question of whether a human embryo is a human individual is addressed by reliance on a clear understanding of the term "potency." The proximate, active potency to autonomously direct the formation of the cells, tissues, organs, and structures required for increasingly mature stages of human life (as witnessed by the observable events of development) is what identifies a human zygote as a human individual. This potency to undergo *development* arises from the human substantial form that is logically and metaphysically prior to the formation of the zygote itself. In cases of monozygotic twinning, a remote, active potency of parts to autonomously direct their own development, not as a part, but as an integrated and independent whole, becomes proximate; that is, a substantial change occurs. This can occur by the separation of a single early blastomere from the embryo or by separation of a group of cells (e.g., the demi-embryo resulting from embryo splitting at the blastocyst stage). Yet in both cases, it is the proximate, active potency for development (i.e., integrated function) that uniquely defines a human individual.

What is the basis of human value? To the question of what makes human beings worthy of human rights there is only a single answer that is consistent with the concepts of justice, liberty, and equality: human beings are human individuals with intrinsic human rights by virtue of their nature as human beings. Thus, all human organisms (human beings) are also human individuals with fundamental human rights.

Summary

This analysis demonstrates that, based on universally accepted scientific criteria, human life begins at the instant of sperm-egg fusion. From this moment forward, the embryo autonomously directs its own development to produce the cells, interactions, and structures necessary for the ongoing process of maturation. Human zygotes exhibit a proximate, active totipotency and are therefore human organisms (i.e., human beings). The preponderance of the scientific evidence strongly suggests that totipotency does not persist beyond the two-cell stage. Yet at all stages in which blastomeres exist that can develop into a mature individual when isolated, such blastomeres possess only a *remote*, active potency to direct their own development; that is, blastomeres are naturally a part of an embryo and must undergo a substantial change to become independent embryos. Totipotency reflects the ability of a zygote to undergo *development*, which minimally requires the ability to produce the first two cell types as part of an integrated spatial and temporal sequence. Twinning, whether it occurs at the two-cell stage or (far more likely) at the blastocyst stage, disrupts an ongoing developmental sequence of a single human organism. Consequently, twinning is an example of asexual human reproduction, where one twin is ontologically identical to the original embryo and one is newly formed at the moment of splitting. At splitting, the cells constituting the newly formed twin undergo a substantial change from being a part to being a whole; that is, a new human substantial form (soul) comes into existence. Our inability to distinguish which twin is the original embryo and which is newly formed is an epistemological, not an ontological, problem. In all cases of twinning, the parents of the original embryo, as the intentional participants in a reproductive act, are the parents of both twins. Similar reasoning applies to the converse process of embryo fusion to form a chimeric individual. Philosophical concerns regarding twinning reflect an inadequate understanding of potency, intention, and individuality. In contrast, scientific confusions about the embryo largely reflect the culture of science, which incentivizes a strictly utilitarian view of the embryo. Finally, human embryos are human individuals, by virtue of the same feature that makes them human organisms—that is, the presence of a human organizing principle or substantial form. Assigning human value based on what humans intrinsically are, rather than on some aspect of human form or function, is the only view of human rights that is consistent with the principles of justice, liberty, and equality.

GLOSSARY

Note: Wherever possible, definitions are taken from *Merriam-Webster's Medical Dictionary*, with minor modifications and/or additions introduced for clarity shown in italics. For terms not included in this dictionary, definitions provided by the author are also indicated by italics.

ANT:	*Altered nuclear transfer; a modification of SCNT/ Cloning designed to generate a nonembryo.*
amnion:	A thin membrane forming a closed sac about the embryos and fetuses of reptiles, birds, and mammals and containing the amniotic fluid—*called also amniotic sac.*
Androgenesis (androgenote):	Development of an embryo containing only paternal chromosomes due to failure of the egg nucleus to participate in fertilization.
blastocyst:	The modified blastula of a placental mammal; an early metazoan embryo typically having the form of a hollow fluid-filled rounded cavity bounded by a single layer of cells.
chimera:	An individual, organ, or part containing tissue with two or more genetically distinct populations of cells. *Chimeras are distinct from mosaics because the distinct populations of cells within a chimera arise from different organisms.*

chorion:	The highly vascular outer embryonic membrane that is associated with the allantois in the formation of the placenta.
cytoplast:	*The material or protoplasm within a living cell, excluding the nucleus.*
development:	The process of growth and differentiation by which the potentialities of a zygote, spore, or embryo are realized. *I have defined the ordered formation of the first two distinct committed cell types as the minimum requirement for development.*
diploid:	Having the basic (*haploid*) chromosome number doubled. *Diploid is the normal state for somatic (i.e., body) cells.*
dizygotic twins:	Twins derived from two ova; fraternal *or nonidentical* twins.
DNA:	Any of various nucleic acids that are usually the molecular basis of heredity, are constructed of a double helix held together by hydrogen bonds between purine and pyrimidine bases which project inward from two chains containing alternate links of deoxyribose and phosphate, and in eukaryotes are localized chiefly in cell nuclei—also called deoxyribonucleic acid.
embryo:	An animal in the early stages of growth and differentiation that are characterized by cleavage, the laying down of fundamental tissues, and the formation of primitive organs and organ systems; especially the developing human individual from the time of *fertilization*[1] to the end of the eighth week

---

1. The definition given by *Merriam-Webster's* indicates "implantation," yet this is inconsistent with the evidence presented here and with the definition of "implantation" (see below). Moreover, this definition is a recent alteration that appears to be politically motivated; see Gacek, "Conceiving Pregnancy."

after conception *(cleavage commences immediately after fertilization to produce the two-cell embryo).*

enucleated egg: An egg or oocyte lacking a nucleus.

epigenetic: Relating to, being, or involving a modification in gene expression that is independent of the DNA sequence of a gene.

fertilization: The process of union of two gametes whereby the somatic chromosome number is restored and the development of a new individual is initiated.

gamete: A mature male or female germ cell *(sperm or egg)* usually possessing a haploid chromosome set and capable of initiating formation of a new diploid individual by fusion with a gamete of the opposite sex—called also sex cell.

gastrulation: The process of becoming or of forming a gastrula, an early metazoan embryo in which the ectoderm, mesoderm, and endoderm are established either by invagination of the blastula (as in fish and amphibians) to form a multilayered cellular cup with a blastopore opening into the archenteron or (as in reptiles, birds, and mammals) by differentiation of the upper layer of the blastodisc into the ectoderm and the lower layer into the endoderm and by the inward migration of cells through the primitive streak to form the mesoderm. *In gastrulation, the main axes of the body are established and three early embryonic tissues are formed (ectoderm, mesoderm, and endoderm) that will subsequently produce the structures of the mature body. In human embryos, identical twinning is no longer possible after gastrulation begins at fourteen days after sperm-egg fusion.*

gene: A specific sequence of nucleotides in DNA or RNA that is located usually on a chromosome and that is the functional unit of inheritance controlling the transmission and expression of one or more traits by specifying the structure of a particular polypeptide and especially a protein or controlling the function of other genetic material—also called determinant, determiner, factor.

genome:

One haploid set of chromosomes with the genes they contain.

haploid:

Having the gametic number of chromosomes or half the number characteristic of somatic cells.

hydatidiform mole:

A mass in the uterus that consists of enlarged edematous degenerated chorionic villi growing in clusters resembling grapes, that typically develops following fertilization of an enucleate egg, and that may or may not contain fetal tissue. *A complete mole is a haploid or diploid entity containing only paternally derived chromosomes, and a partial mole is a triploid entity containing a preponderance of paternally derived chromosomes.*

implantation:

In placental mammals: the process of attachment of the embryo to the maternal uterine wall—called also nidation.

inner cell mass:

(ICM) The portion of the blastocyst of a mammalian embryo that is destined to become the *structures of the postnatal body.*

meiosis:

The cellular process that results in the number of chromosomes in gamete-producing cells being reduced to one half and that involves a reduction division in which one of each pair of homologous chromosomes passes to each daughter cell and a mitotic division.

membrane:

A semipermeable limiting layer of cell protoplasm consisting of a fluid phospholipid bilayer with intercalated proteins—called also cell membrane, plasmalemma.

mitosis:
A process that takes place in the nucleus of a dividing cell, involves typically a series of steps consisting of prophase, metaphase, anaphase, and telophase, and results in the formation of two new nuclei, each having the same number of chromosomes as the parent nucleus.

monozygotic twins: Twins derived from a single egg; *identical twins.*

morula:
A globular solid mass of blastomeres formed by cleavage of a zygote that typically precedes the blastula (*blastocyst for human embryos*).

mosaic:
An organism or one of its parts composed of cells of more than one genotype. *Mosaics are distinct from chimeras because the genetic differences within a mosaic arise by alterations of a single, preexisting genome.*

nucleus:
A cellular organelle of eukaryotes that is essential to cell functions (as reproduction and protein synthesis), is composed of nuclear sap and a nucleoprotein-rich network from which chromosomes and nucleoli arise, and is enclosed in a definite membrane.

ooplast:
A cytoplast derived from an oocyte or egg cell.

organism:
An individual constituted to carry on the activities of life by means of organs separate in function but mutually dependent: a living being.

ovum
(oocyte, egg):
A female gamete—especially a mature egg that has undergone reduction, is ready for fertilization, and takes the form of a relatively large inactive gamete providing a comparatively great amount of reserve material and contributing most of the cytoplasm of the zygote.

parthenote: Development in which the embryo contains only maternal chromosomes due to activation of an egg by a sperm that degenerates without fusing with the egg nucleus *or due to experimental activation; also, gynogenesis or parthenogenesis.*

placenta: The vascular organ in mammals that unites the fetus to the maternal uterus and mediates its metabolic exchanges through a more or less intimate association of uterine mucosal with chorionic and usually allantoic tissues permitting exchange of material by diffusion between the maternal and fetal vascular systems but without direct contact between maternal and fetal blood and typically involving the interlocking of fingerlike vascular chorionic villi with corresponding modified areas of the uterine mucosa.

plenipotent: *Able to produce all of the cell types derived from both inner cell mass and trophectoderm, but not able to organize them into a coherent body plan.*

pluripotent: Not fixed as to developmental potentialities; especially: capable of differentiating into one of many cell types; *typically used to refer to stem cells that are able to produce all of the cell types of the mature body, but not those derived from trophectoderm.*

polyspermy: The entrance of several spermatozoa into one egg.

pronucleus: The haploid nucleus of a male or female gamete (as an egg or sperm) up to the time of fusion with that of another gamete in fertilization.

reprogramming: *Altering the epigenetic state of a nucleus such that it enters into a new developmental state; e.g., during cloning, a body cell nucleus is reprogrammed by factors within an oocyte to enter into a state similar to that of a zygotic nucleus and capable of supporting a normal pattern of embryonic development.*

SCNT/Cloning:	*Somatic cell nuclear transfer (SCNT): transplanting nuclei from body (i.e., somatic) cells to enucleated eggs.*
somatic:	Of, relating to, or affecting the body especially as distinguished from the germplasm or psyche.
sperm (spermatozoon):	*A sperm cell*; a motile male gamete of an animal usually with rounded or elongated head and a long posterior flagellum.
teratoma:	A tumor derived from more than one embryonic layer and made up of a heterogeneous mixture of tissues (as epithelium, bone, cartilage, or muscle).
tetraploid:	Having or being a chromosome number four times the monoploid (*i.e., haploid*) number.
totipotent:	Capable of developing into a complete organism.
triploid:	Having or being a chromosome number three times the monoploid (*i.e., haploid*) number.
trophectoderm:	*(TE)* The outer layer of the mammalian blastocyst after differentiation of the ectoderm, mesoderm, and endoderm when the outer layer is continuous with the ectoderm of the embryo. *Trophectoderm gives rise predominantly to the placenta.*
zona pellucida:	The transparent, more-or-less elastic noncellular glycoprotein outer layer or envelope of a mammalian ovum.
zygote:	A cell formed by the union of two gametes; broadly, the developing individual produced from such a cell.

Here is a compilation of quotations regarding when human life begins, taken from medical textbooks on human embryology or reproduction and from the peer-reviewed scientific literature. I have added emphasis (*italics*) and offered clarifications in brackets. As discussed above, human embryos are human organisms (i.e., human beings), and only organisms *undergo development*. "Zygote" is the scientific name for a one-cell embryo.

I. Medical Textbooks

1. Keith L. Moore, T. V. N. Persaud, and Mark G. Torchia, *The Developing Human: Clinically Oriented Embryology*, 10th ed. (Philadelphia: Saunders, 2016), 11.

   "Human development begins at fertilization, when a sperm fuses with an oocyte to form a single cell, the zygote. This highly specialized, *totipotent cell* (capable of giving rise to any cell type) marks *the beginning of each of us as a unique individual*."

2. Gary C. Schoenwolf, Steven B. Bleyl, Philip R. Brauer, and Philippa H. Francis-West, *Larsen's Human Embryology*, 5th ed. (Philadelphia: Elsevier Saunders, 2015), 2.

   "*All of us were once human embryos*, so the study of human embryology is the study of our own prenatal origins and experiences."

3. Gary C. Schoenwolf, Steven B. Bleyl, Philip R. Brauer, and Philippa H. Francis-West, *Larsen's Human Embryology*, 5th ed. (Philadelphia: Elsevier Saunders, 2015), 14.

"Fertilization, the uniting of egg and sperm, takes place in the oviduct. After the oocyte finishes meiosis, the paternal and maternal chromosomes come together, resulting in the formation of a zygote containing a single diploid nucleus. *Embryonic development is considered to begin at this point.*"

4. Richard Jones and Kristen H. Lopez, *Human Reproductive Biology*, 4th ed. (Waltham, MA: Elsevier Academic, 2014), 169.

"*The fertilized egg (zygote) is the beginning of a new diploid individual.*"

5. Keith L. Moore, *Before We Are Born: Essentials of Embryology*, 7th ed. (Philadelphia: Saunders, 2008), 2.

"[The zygote], formed by the union of an oocyte and a sperm, *is the beginning of a new human being.*"

6. Keith L. Moore, *Before We Are Born: Essentials of Embryology*, 9th ed. (Philadelphia: Saunders, 2016), 1.

"*Human development begins at fertilization* when an oocyte (ovum) from a female is fertilized by a sperm (spermatozoon) from a male. . . . Embryology is concerned with the origin and development of a human being from a zygote to birth."

7. T. W. Sadler, *Langman's Medical Embryology*, 10th ed. (Philadelphia: Lippincott Williams and Wilkins, 2006), 11.

"*Development begins with fertilization*, the process by which the male gamete, the sperm, and the female gamete, the oocyte, unite to give rise to a zygote."

8. T. W. Sadler, *Langman's Medical Embryology*, 13th ed. (Philadelphia: Lippincott Williams and Wilkins, 2015), 42.

"The main results of fertilization are as follows: Restoration of the diploid number of chromosomes, half from the father half from the mother. Hence, the zygote contains a new combination of chromosomes different from both parents. Determination of the sex *of the new individual.* An

X-carrying sperm produces a female (XX) embryo and a Y-carrying sperm produces a male (XY) embryo. Therefore, the chromosomal sex of the embryo is determined at fertilization."

9. Ronald W. Dudek, *Embryology*, 4th ed. (Philadelphia: Lippincott Williams and Wilkins, 2008), 1.

"Sexual reproduction occurs when female and male gametes (oocyte and spermatozoon, respectively) *unite at fertilization.*"

10. Ronan O'Rahilly and Fabiola Miller, *Human Embryology and Teratology*, 3rd ed. (New York: Wiley-Liss, 2001), 8.

"Although life is a continuous process, fertilization . . . is a critical landmark because, under ordinary circumstances, *a new genetically distinct human organism is formed when the chromosomes of the male and female pronuclei blend in the oocyte.*"

11. Bruce M. Carlson, *Human Embryology and Developmental Biology*, 5th ed. (Philadelphia: Elsevier Saunders, 2014), 2.

"*Human pregnancy begins with the fusion of an egg and a sperm* within the female reproductive tract."

II. Peer-Reviewed Scientific Literature, 2001 to the Present (Chronological Order)

1. Paul M. Wassarman, Luca Jovine, and Eveline S. Litscher, "A Profile of Fertilization in Mammals," *Nature Cell Biology* 3, no. 2 (February 2001): E59.

"When mammalian eggs and sperm come into contact in the female oviduct, a series of steps is set in motion that can lead to fertilization and ultimately to development of *new individuals.*"

2. Paul Primakoff and Diana Gold Myles, "Penetration, Adhesion, and Fusion in Mammalian Sperm-Egg Interaction," *Science* 296, no. 5576 (June 2002): 2183.

"Fertilization is the sum of the cellular mechanisms that pass the genome from one generation to the next and *initiate development of a new organism.*"

3. Linda L. Runft, Laurinda A. Jaffe, and Lisa M. Mehlmann. "Egg Activation at Fertilization: Where It All Begins," *Developmental Biology* 245, no. 2 (May 2002): 237.

[Note the use of the phrase] "Where It All Begins."

4. Luigia Santella, Emanuela Ercolano, and Gilda A. Nusco. "The Cell Cycle: A New Entry in the Field of Ca2+ Signaling," *Cellular and Molecular Life Sciences* 62, no. 21 (November 2005): 2405.

"Ca2+ signaling plays a crucial role in virtually all cellular processes, from the *origin of new life at fertilization* to the end of life when cells die."

5. Naokazu Inoue, Masahito Ikawa, Ayako Isotani, and Masaru Okabe, "The Immunoglobulin Superfamily Protein Izumo Is Required for Sperm to Fuse with Eggs," *Nature* 434, no. 7030 (March 2005): 234.

"Representing the 60 trillion cells that build a human body, a sperm and an egg meet, recognize each other, and *fuse to form a new generation of life.*"

6. Ken-ichi Sato, Yasuo Fukami, and Bradley J. Stith, "Signal Transduction Pathways Leading to Ca2+ Release in a Vertebrate Model System: Lessons from Xenopus Eggs," *Seminars in Cell & Developmental Biology* 17, no. 2 (April 2006): 285.

"At fertilization, eggs unite with sperm to initiate developmental programs that give rise to development of the embryo. Defining the molecular mechanism of this fundamental process at *the beginning of* life has been a key question in cell and developmental biology."

7. Meital Oren-Suissa and Benjamin Podbilewicz, "Cell Fusion during Development," *Trends in Cell Biology* 17, no. 11 (November 2007): 537.

"Most readers of this review originated from a sperm-egg fusion event."

8. Khaled Machaca, "Ca2+ Signaling Differentiation during Oocyte Maturation," *Journal of Cellular Physiology* 213, no. 2 (November 2007): 331.

"Oocyte maturation is an essential cellular differentiation pathway that prepares the egg for activation at fertilization leading to the *initiation of embryogenesis.*"

9. Naokazu Inoue and Masaru Okabe, "Sperm-Egg Fusion Assay in Mammals," *Methods in Molecular Biology* 475 (2008): 335.

"As representatives of the 60 trillion cells that make a human body, a sperm and an egg meet, recognize each other, and fuse to *create a new generation.*"

10. Tammy F. Wu and Diana S. Chu, "Sperm Chromatin: Fertile Grounds for Proteomic Discovery of Clinical Tools," *Molecular & Cellular Proteomics* 7, no. 10 (October 2008): 1876.

"Sperm are remarkably complex cells with a singularly important mission: to deliver paternal DNA and its associated factors to the oocyte *to start a new life.*"

11. Alexei V. Evsikov and Caralina Marín de Evsikova, "Gene Expression during the Oocyte-to-Embryo Transition in Mammals," *Molecular Reproduction and Development* 76, no. 9 (September 2009): 805.

"The seminal question in modern developmental biology is *the origins of new life* arising from the unification of sperm and egg."

12. Yasunori Okada, Kazuo Yamagata, Kwonho Hong, Teruhiko Wakayama, and Yi Zhang, "A Role for Elongator in Zygotic Paternal Genome Demethylation," *Nature* 463, no. 7280 (January 2010): 554.

"The life cycle of mammals begins when a sperm enters an egg."

13. Matthew R. Marcello and Andrew W. Singson, "Fertilization and the Oocyte-to-Embryo Transition in C. Elegans," *BMB Reports* 43, no. 6 (June 2010): 389.

"Fertilization is a complex process comprised of numerous steps. During fertilization, two highly specialized and differentiated cells (sperm and egg) fuse and subsequently trigger the *development of an embryo* from a quiescent, arrested oocyte."

14. Hana Robson Marsden, Itsuro Tomatsu, and Alexander Kros, "Model Systems for Membrane Fusion," *Chemical Society Reviews* 40, no. 3 (March 2011): 1572.

"The fusion of sperm and egg membranes *initiates the life* of a sexually reproducing organism."

15. Alberto Darszon, Takuya Nishigaki, Carmen Beltran, and Claudia Lydia Treviño, "Calcium Channels in the Development, Maturation, and Function of Spermatozoa," *Physiological Reviews* 91, no. 4 (October 2011): 1305.

"A proper dialogue between spermatozoa and the egg is essential for conception of a new individual in sexually reproducing animals. Ca(2+) is crucial in orchestrating this unique event *leading to a new life.*"

16. Ting Ting Sun, Chin Man Chung, and Hsiao Chang Chan, "Acrosome Reaction in the Cumulus Oophorus Revisited: Involvement of a Novel Sperm-Released Factor NYD-SP8," *Protein & Cell* 2, no. 2 (February 2011): 92.

"Fertilization is a process involving multiple steps that lead to the final fusion of one sperm and oocyte to *form the zygote.*"

17. Michail Nomikos, Karl Swann, and F. Anthony Lai, "Starting a New Life: Sperm PLC-zeta Mobilizes the Ca2+ Signal That Induces Egg Activation and Embryo Development; An Essential Phospholipase C with Implications for Male Infertility," *BioEssays: News and Reviews in Molecular, Cellular and Developmental Biology* 34, no. 2 (February 2012): 126.

[Note the use of the phrase] "Starting a New Life."

18. Janetti R. Signorelli, E. Santana Diaz, and Patricio Morales, "Kinases, Phosphatases and Proteases during Sperm Capacitation," *Cell and Tissue Research* 349, no. 3 (September 2012): 765.

"Fertilization is the process by which male and female haploid gametes (sperm and egg) unite to *produce a genetically distinct individual.*"

19. Pilar Coy, Francisco Alberto García-Vázquez, Pablo E. Visconti, and Manuel Avilés, "Roles of the Oviduct in Mammalian Fertilization," *Reproduction* 144, no. 6 (December 2012): 649.

"The oviduct or Fallopian tube *is the anatomical region where every new life begins* in mammalian species. After a long journey, the spermatozoa meet the oocyte in the specific site of the oviduct named ampulla, and fertilization takes place."

20. Dessie Salilew-Wondim, Karl Schellander, Michael Hoelker, and Dawit Tesfaye, "Oviductal, Endometrial and Embryonic Gene Expression Patterns as Molecular Clues for Pregnancy Establishment," *Animal Reproduction Science* 134, nos. 1–2 (September 2012): 9.

"In higher animals, the *beginning of new life* and transfer of genetic material to the next generation occurs in the oviduct when two distinct gametes [*sic*] cells unite resulting in the formation of a zygote."

21. Janice P. Evans, "Sperm-Egg Interaction," *Annual Review of Physiology* 74 (2012): 477.

"A crucial step of fertilization is the sperm-egg interaction that allows the *two gametes to fuse and create the zygote.*"

22. Takuya Wakai, Veerle Vanderheyden, Suk Bong Yoon, Banyoon Cheon, Nan Zhang, Jan B. Parys, and Rafael A. Fissore, "Regulation of Inositol 1,4,5-Trisphosphate Receptor Function during Mouse Oocyte Maturation," *Journal of Cellular Physiology* 227, no. 2 (February 2012): 705.

"At the time of fertilization, an increase in the intracellular $Ca(2+)$ concentration ($[Ca(2+)](i)$) underlies egg activation and *initiation of development* in all species studied to date."

23. M. R. Marcello, G. Singaravelu, and A. Singson, "Fertilization," *Advances in Experimental Medical Biology* 757 (2013): 321.

"Fertilization—the fusion of gametes to *produce a new organism*—is the culmination of a multitude of intricately regulated cellular processes."

24. Scott Robertson and Rueyling Lin, "The Oocyte-to-Embryo Transition," *Advances in Experimental Medical Biology* 757 (2013): 351.

"The oocyte-to-embryo transition refers to the process whereby a fully grown, relatively quiescent oocyte undergoes maturation, fertilization, and is *converted into a developmentally active, mitotically dividing embryo*, arguably one of the most dramatic transitions in biology."

25. Nancy Nader, Rashmi P. Kulkarni, Maya Dib, and Khaled Machaca, "How to Make a Good Egg! The Need for Remodeling of Oocyte Ca(2+) Signaling to Mediate the Egg-to-Embryo Transition," *Cell Calcium* 53, no. 1 (January 2013): 41.

"The egg-to-embryo transition marks *the initiation of multicellular organismal development* and is mediated by a specialized Ca(2+) transient at fertilization."

26. Atsushi Asano, Jacquelyn L. Nelson-Harrington, and Alexander J. Travis, "Membrane Rafts Regulate Phospholipase B Activation in Murine Sperm," *Communicative & Integrative Biology* 6, no. 6 (November 2013): e27362.

"It is intuitive that *fertilization—the start of life*—involves communication between a sperm cell and an egg."

27. Enrica Bianchi, Brendan Doe, David A. Goulding, and Gavin J. Wright, "Juno Is the Egg Izumo Receptor and Is Essential for Mammalian Fertilisation," *Nature* 508, no. 7497 (April 2014): 483.

"Fertilization occurs when *sperm and egg recognize each other and fuse to form a new, genetically distinct organism.*"

28. Diego Lorenzetti, Christophe Poirier, Ming Zhao, Paul Overbeek, Wilbur R. Harrison, and Colin Edward Bishop, "A Transgenic Insertion on Mouse Chromosome 17 Inactivates a Novel Immunoglobulin Superfamily Gene Potentially Involved in Sperm–Egg Fusion." *Mammalian Genome* 25, nos. 3–4 (April 2013): 141.

"Fertilization is the process that *leads to the formation of a diploid zygote* from two haploid gametes."

29. Allison Tscherner, Graham C. Gilchrist, Natasha Smith, Patrick Blondin, Daniel J. Gillis, and Jonathan Lamarre, "MicroRNA-34

Family Expression in Bovine Gametes and Preimplantation Embryos," *Reproductive Biology and Endocrinology* 12 (September 2014): 85.

"In sexually reproducing organisms, *embryogenesis begins with the fusion of two haploid gametes.*"

30. Enrica Bianchi and Gavin J. Wright, "Cross-Species Fertilization: The Hamster Egg Receptor, Juno, Binds the Human Sperm Ligand, Izumo1," *Philosophical Transactions of the Royal Society of London; Series B, Biological Sciences* 370, no. 1661 (February 2015): 20140101.

"Fertilization is the culminating event in sexual reproduction and requires the recognition and *fusion of the haploid sperm and egg to form a new diploid organism.*"

31. Martin Graham Wilding, Gianfranco Coppola, Francesco De Icco, Laura Arenare, Loredana di Matteo, and Brian Dale, "Maternal Non-Mendelian Inheritance of a Reduced Lifespan? A Hypothesis," *Journal of Assisted Reproduction and Genetics* 31, no. 6 (June 2014): 637.

"Since *a new individual is derived from the fusion of a single sperm and egg,* we tested."

32. Bart M. Gadella and Arjan Boerke, "An Update on Post-Ejaculatory Remodeling of the Sperm Surface before Mammalian Fertilization," *Theriogenology* 85, no. 1 (January 2016): 113.

"*The fusion of a sperm with an oocyte to form new life* is a highly regulated event."

33. Junaid Kashir, Michail Nomikos, Karl Swann, and F. Anthony Lai, "PLCζ or PAWP: Revisiting the Putative Mammalian Sperm Factor That Triggers Egg Activation and Embryogenesis," *Molecular Human Reproduction* 21, no. 5 (May 2015): 383.

"In mammals, egg activation is initiated by multiple cytosolic Ca(2+) transients (Ca(2+) oscillations) that are triggered following delivery of a putative sperm factor from the fertilizing sperm. The identity of this 'sperm factor' thus holds much significance, *not only as a vital component in creating a new life*, but also for its potential therapeutic and diagnostic value in human infertility."

34. Robert J. Norman, "2015 RANZCOG Arthur Wilson Memorial Oration 'From Little Things, Big Things Grow: The Importance of Periconception Medicine,'" *The Australian & New Zealand Journal of Obstetrics & Gynaecology* 55, no. 6 (December 2015): 535.

    "The time of our conception is when we are most vulnerable to survival and growing as a healthy human being."

35. Lena Willkomm and Wilhelm Bloch, "State of the Art in Cell-Cell Fusion," *Methods in Molecular Biology* 1313 (2015): 1.

    "Mammalian life begins with a cell-cell fusion event, i.e. the fusion of the spermatozoid with the oocyte."

36. Chuen Yan Leung and Magdalena Zernicka-Goetz, "Mapping the Journey from Totipotency to Lineage Specification in the Mouse Embryo," *Current Opinion in Genetics & Development* 34 (October 2015): 71.

    "Mammalian life, with all its complexity *comes from a humble beginning of a single fertilized egg cell.*"

37. Patricio Ventura-Juncá, Isabel Irarrázaval, Augusto J. Rolle, Juan I. Rodríguez Gutiérrez, Ricardo D. Moreno, and Manuel J. Pires dos Santos, "In Vitro Fertilization (IVF) in Mammals: Epigenetic and Developmental Alterations; Scientific and Bioethical Implications for IVF in Humans," *Biological Research* 48 (December 2015): 68.

    "The advent of in vitro fertilization (IVF) in animals and humans implies an extraordinary change in the environment where the beginning of a new organism takes place."

38. Enrica Bianchi and Gavin J. Wright, "Sperm Meets Egg: The Genetics of Mammalian Fertilization," *Annual Review of Genetics* 50 (November 2016): 93.

"Fertilization is the culminating event of sexual reproduction, which involves the union of the sperm and egg to form a single, genetically distinct organism."

39. Lei Guo, Li Si Jiang, Ying Zhang, Xiu-Li Lu, Qi Xie, Dolf Weijers, and Chun-Ming Liu, "The Anaphase-Promoting Complex Initiates Zygote Division in Arabidopsis through Degradation of Cyclin B1," *The Plant Journal: For Cell and Molecular Biology* 86, no. 2 (April 2016): 161.

"As the start of a new life cycle, activation of the first division of the zygote is a critical event in both plants and animals."

40. Bikem Soygur and Leyla Sati, "The Role of Syncytins in Human Reproduction and Reproductive Organ Cancers," *Reproduction* 152, no. 5 (November 2016): R167.

"Human life begins with sperm and oocyte fusion."

41. Diana Chu, "Parental Control Begins at the Beginning," *Genetics* 204, no. 4 (December 2016): 1377.

"New parents anticipate their job begins at birth. Little do they know *they have been exerting control within the baby's first cell since fertilization.*"

42. Warif El Yakoubi and Katja Wassmann, "Meiotic Divisions: No Place for Gender Equality," *Advances in Experimental Medicine and Biology* 1002 (2017): 1.

"In multicellular organisms *the fusion of two gametes with a haploid set of chromosomes leads to the formation of the zygote, the first cell of the embryo.*"

43. Zoltán Macháty, "Signal Transduction in Mammalian Oocytes during Fertilization," *Cell and Tissue Research* 363, no. 1 (January 2016): 169.

"Mammalian embryo development begins when the fertilizing sperm triggers a series of elevations in the oocyte's intracellular free Ca(2+) concentration."

44. Tomoya Fukui, Lei Guo, Xiu-Li Hou, Hua-Qin Gong, and Chun-Ming Liu, "ZYGOTE-ARREST 3 That Encodes the tRNA Ligase Is

Essential for Zygote Division in Arabidopsis," *Journal of Integrative Plant Biology* 59, no. 9 (September 2017): 680.

"In sexual organisms, *division of the zygote initiates a new life cycle.*"

45. Katherine Lynn Wozniak, Guillermina Maria Luque, and Soo Hyun Ahn, "When Sperm Meets Egg: The Spark of New Life," *Molecular Reproduction and Development* 85, no. 1 (January 2018): 5.

[Note the use of the phrase] "The Spark of New Life."

46. Catherine Rollo, Yexia Li, Xing Liang Jin, and Caitlin G. O'Neill, "Histone 3 Lysine 9 Acetylation Is a Biomarker of the Effects of Culture on Zygotes," *Reproduction* 154, no. 4 (October 2017): 375.

"Fertilisation triggers a round of chromatin remodelling that prepares the genome for the first round of transcription from the *new embryonic genome.*"

47. Muhammad Saad Ahmed, Sana Ikram, Nousheen Bibi, and Asif Mir, "Hutchinson–Gilford Progeria Syndrome: A Premature Aging Disease," *Molecular Neurobiology* 55, no. 5 (May 2017): 4417.

"Aging is a developmental process *that begins with fertilization* and ends up with death involving a lot of environmental and genetic factors."

48. Zoltán Macháty, Andrew R. Miller, and Lu Zhang, "Egg Activation at Fertilization," *Advances in Experimental Medicine and Biology* 953 (2017): 1.

"Fertilization is the union of gametes to *initiate development of a new individual.*"

49. Joanna W. Jachowicz, Xinyang Bing, Julien Pontabry, Ana Bošković, Oliver J. Rando, and Maria-Elena Torres-Padilla, "LINE-1 Activation after Fertilization Regulates Global Chromatin Accessibility in the Early Mouse Embryo," *Nature Genetics* 49, no. 10 (October 2017): 1502.

[referring to events in the zygote] "Our data suggest that activation of LINE-1 regulates global chromatin accessibility *at the beginning of development* and indicate that retrotransposon activation is integral to the developmental program."

50. Itai Gat and Raoul Orvieto, "'This Is Where It All Started'—The Pivotal Role of PLCζ within the Sophisticated Process of Mammalian Reproduction: A Systemic Review," *Basic and Clinical Andrology* 27 (2017): 9.

"At the end of oogenesis and spermatogenesis, both haploid gametes contain a single set of chromosomes ready *to form the zygote, the first cell of the newly developing individual.*"

51. Yuki Okada and Kosuke Yamaguchi, "Epigenetic Modifications and Reprogramming in Paternal Pronucleus: Sperm, Preimplantation Embryo, and Beyond," *Cellular and Molecular Life Sciences* 74, no. 11 (June 2017): 1957.

"*Pronuclear/zygotic stage is the very first stage of life.*"

52. Alaa Hachem, Jonathan M. Godwin, Margarida Ruas, Hoi Chang Lee, Minerva Ferrer Buitrago, Goli Ardestani, Andrew R. Bassett, et al., "PLCζ Is the Physiological Trigger of the Ca2+ Oscillations That Induce Embryogenesis in Mammals but Conception Can Occur in Its Absence," *Development* 144, no. 16 (August 2017): 2914.

"*Activation of the egg by the sperm is the first, vital stage of embryogenesis.*"

53. I. Christy Raj, Hamed Sadat Al Hosseini, Elisa Dioguardi, Kaoru Nishimura, Ling Bo Han, Alessandra Villa, Daniele de Sanctis, and Luca Jovine, "Structural Basis of Egg Coat-Sperm Recognition at Fertilization," *Cell* 169, no. 7 (June 2017): 1315.

"Recognition between sperm and the egg surface *marks the beginning of life* in all sexually reproducing organisms."

54. Ryan J. Gleason, Amit Anand, Toshie Kai, and Xin Chen, "Protecting and Diversifying the Germline," *Genetics* 208, no. 2 (February 2018): 435.

"*Restarting a new life cycle upon fertilization is a unique challenge faced by gametes.*"

55. Antonietta Salustri, Luisa Campagnolo, Francesca Gioia Klinger, and Antonella Camaioni, "Molecular Organization and Mechanical Properties of the Hyaluronan Matrix Surrounding the Mammalian Oocyte," *Matrix Biology: Journal of the International Society for Matrix Biology* (February 2018): pii: S0945-053X(18)30025.

> *"Successful ovulation and oocyte fertilization are essential prerequisites for the beginning of life in sexually reproducing animals."*

56. Serafín Pérez-Cerezales, Priscila Ramos-Ibeas, Omar Salvador Acuña, Manuel Avilés, Pilar Coy, Dimitrios Rizos, and Alfonso Gutiérrez-Adán, "The Oviduct: From Sperm Selection to the Epigenetic Landscape of the Embryo," *Biology of Reproduction* 98, no. 3 (March 2018): 262.

> *"The mammalian oviduct is the place where life begins as it is the site of fertilization and preimplantation embryo development."*

57. Shinnosuke Suzuki and Naojiro Minami, "CHD1 Controls Cell Lineage Specification Through Zygotic Genome Activation," *Advances in Anatomy, Embryology, and Cell Biology* 229 (2018): 15.

> *"Life begins with the encounter of eggs and spermatozoa."*

58. John Parrington, Christophe Arnoult, and Rafael A. Fissore, "The Eggstraordinary Story of How Life Begins," *Molecular Reproduction and Development* 86, no. 1 (January 2018): 4.

> "This molecule [provided by the sperm at sperm-egg fusion] was proposed to represent the sperm factor responsible for the initiation of calcium (Ca^{2+}) oscillations *required for egg activation and embryo development* in mammals."

59. Shang Wang and Irina V. Larina, "In Vivo Imaging of the Mouse Reproductive Organs, Embryo Transfer, and Oviduct Cilia Dynamics Using Optical Coherence Tomography," *Methods in Molecular Biology* 1752 (2018): 53.

> "The oviduct (or fallopian tube) serves as the site where a number of major reproductive events occur *for the start of a new life* in mammals."

60. Jinsong Liu, "The Dualistic Origin of Human Tumors," *Seminars in Cancer Biology* 53 (December 2018): 1.

"Life starts with a zygote, which is formed by the *fusion of a haploid sperm and egg*."

NOTES

Introduction

1. See, for example, Gilbert, Tyler, and Zackin, *Bioethics and the New Embryology*, 31.

2. See, for example, Rankin, "Can One Be Two?"; Griniezakis and Symeonides, "Twin Conception"; St. John, "And on the Fourteenth Day."

3. Grobstein, "External Human Fertilization."

4. Grobstein, Flower, and Mendeloff, "External Human Fertilization."

5. Kiernan, "Pre-embryos"; "IVF Remains in Legal Limbo."

6. McLaren, "Pre-embryos?"; West et al., "Sexing the Human Pre-embryo"; Shaw et al., "Attempts to Stimulate"; Watt et al., "Trisomy 1"; Yu and Chan, "Effects of Cadmium."

7. Grobstein, "Biological Characteristics of the Preembryo," 346.

8. Grobstein, "Biological Characteristics of the Preembryo," 346.

9. Grobstein, "Biological Characteristics of the Preembryo," 346.

10. Ford, *When Did I Begin?*

11. Warnock, *Report of the Committee of Inquiry.*

12. Ferrer Colomer and Pastor, "Preembryo's Short Lifetime"; Vivanco et al., "Bibliometric Analysis."

13. Gilbert, Tyler, and Zackin, *Bioethics and the New Embryology*, 3.

14. Hallgrímsson and Hall, *Epigenetics.*

15. Price Foley, *Law of Life and Death.*

16. Interestingly, British scientists have recently argued to extend the permissible period for embryo experimentation beyond the fourteen-day limit. See, Cook, "UK Scientists to Push."

17. Thomson et al., "Embryonic Stem Cell Lines."

18. Tachibana et al., "Human Embryonic Stem Cells."

19. Tachibana et al., "Towards Germline."

20. Liang et al., "CRISPR/Cas9-Mediated."

21. Han et al., "Forebrain Engraftment."

22. For example, embryonic stem cell research has been conducted for nearly three decades in animal models, and for over twenty years using human cells, without resulting in medical treatments for human disease. See Condic, "The Basics about Stem Cells"; Condic and Rao, "Alternative Sources of Pluripotent Stem Cells."

23. "Embryology."

Chapter 1

1. Takahashi et al., "Induction of Pluripotent Stem Cells."

2. A minor caveat to this conclusion involves *cyclic or transient changes* in cell composition/behavior that are followed by a return to the initial cellular state (for example, molecular changes associated with the mitotic cycle or in response to a chemical signal). Such changes do not signify the formation of a new cell, but rather entry of a cell into a specific, transient, cellular state.

3. Satouh et al., "Visualization of the Moment."

4. Hernández and Podbilewicz, "Hallmarks"; Aguilar et al., "Genetic Basis"; Risselada and Grubmüller, "How SNARE Molecules Mediate"; Marsden, Tomatsu, and Kros, "Model Systems."

5. Condic, "When Does Human Life Begin? The Scientific Evidence"; Condic, "When Does Human Life Begin? A Scientific Perspective."

6. Mills, "Egg and I."

7. Miller and Pruss, "Human Organisms."

8. Miller and Pruss, "Human Organisms," 537.

9. Rosslenbroich, "Properties of Life."

10. *Merriam Webster Medical Dictionary.*

11. Condic, "Totipotency." For a discussion of cases in which development does not occur, see Chapter 3 in this volume.

12. Harper and Sengupta, "Preimplantation Genetic Diagnosis."

13. Huang et al., "Primary Unruptured"; Sehgal et al., "Full Term Ovarian Pregnancy."

14. Dahab et al., "Full-Term Extrauterine"; Isah et al., "Abdominal Pregnancy"; Zhang and Sheng, "Full-Term Abdominal Pregnancy"; Xiao et al., "Abdominal Pregnancy"; Badria et al., "Full-Term Viable."

15. Shukla et al., "Primary Hepatic Pregnancy."

16. Condic, "When Does Human Life Begin? The Scientific Evidence"; Condic, "When Does Human Life Begin? A Scientific Perspective."

17. Recent evidence suggests that differences in gene expression and developmental capabilities occur as early as the two cell stage; see Casser et al., "Retrospective Analysis"; Casser et al., "Totipotency Segregates."

18. Antczak and Blerkom, "Oocyte Influences"; Edwards and Hansis, "Initial Differentiation"; Hansis, Grifo, and Krey, "Candidate Lineage Marker"; Jędrusik et al., "Role of Cdx2"; Sun et al., "Differential Expression"; Plachta et al., "Oct4 Kinetics."

19. Torres-Padilla et al., "Histone Arginine Methylation"; Hupalowska et al., "CARM1 and Paraspeckles"; Wang et al., "Asymmetric Expression."

20. Piotrowska-Nitsche et al., "Four-Cell Stage"; Jędrusik et al., "Role of Cdx2."

21. Condic, "When Does Human Life Begin? The Scientific Evidence."

Chapter 2

1. Baertschi and Mauron, "Moral Status Revisited"; Magill and Neaves, "Ontological and Ethical"; Wasserman, "What Qualifies."

2. *Merriam-Webster's Medical Dictionary*.

3. Feigin and Malbon, "OSTM1 Regulates"; Honecker et al., "Germ Cell Lineage."

4. Abad et al., "Reprogramming In Vivo"; Morgani et al., "Totipotent Embryonic Stem."

5. The potency of embryonic stem cells varies quite a bit from species to species. For example, mouse embryonic stem cells are largely restricted to producing cells of the mature mouse body, and are therefore pluripotent. In contrast, human embryonic stem cells are able to produce both the cells of the mature body and the cells of the placenta and embryonic membranes, and are therefore "totipotent" in the second, more limited sense, or "pluripotent."

6. For discussion of situations in which development is only partially completed or is abnormal, see chapter 3.

7. Condic, "Preimplantation Stages"; Condic, "When Does Human Life Begin? A Scientific Perspective"; Condic, "When Does Human Life Begin? The Scientific Evidence"; Condic, "Totipotency"; Ingebrigtsen, "Studies upon the Characteristics"; Murray, "Physiological Ontogeny."

8. Condic, "Totipotency."

9. Condic, "Totipotency," 787.

10. Paepe et al., "Totipotency and Lineage"; Wu, Lei, and Schöler, "Totipotency."

11. The most common approach is known as "tetraploid complementation," where the first two cells of a normal two-cell embryo are fused to generate a single cell with twice the normal amount of DNA—that is, a tetraploid embryo. Such manipulated embryos typically develop abnormally, because the tetraploid state is not generally compatible with development of the inner cell mass. However, if normal, diploid blastomeres or stem cells are injected into a tetraploid embryo, it is "rescued" (i.e., the defect is "complemented"), and the embryo resumes normal development, with the injected cells producing most of the postnatal body. More sophisticated manipulations to accomplish the same goal are also possible; see, for example, Zhang et al., "Individual Blastomeres."

12. There have been cases of tetraploid humans surviving to live-birth and beyond. See Guć-Šćekić et al., "Tetraploidy"; Nakamura et al., "Tetraploid Live-Born Neonate."

13. For a more detailed discussion of tetraploid complementation assays and how they have been misinterpreted, see Condic, "Totipotency," 798–800.

14. Katayama, Ellersieck, and Roberts, "Development of Monozygotic"; Rossant, "Postimplantation Development"; Tarkowski and Wróblewska, "Development of Blastomeres."

15. Casser et al., "Totipotency Segregates."

16. It is scientifically impossible to prove a negative; i.e., failure to demonstrate totipotency in a particular blastomere does not "prove" the cell is not totipotent, because failure to obtain this result could always be due to limitations in experimental technique or observational resolution. One can only conclude that totipotency was not observed using the methods employed.

17. Tarkowski, Ozdzenski, and Czołowska, "How Many Blastomeres"; Velde et al., "Four Blastomeres."

18. Johnson et al., "Production of Four."

19. Katayama, Ellersieck, and Roberts, "Development of Monozygotic"; Rossant, "Postimplantation Development"; Tarkowski and Wróblewska, "Development of Blastomeres."

20. Saito and Niemann, "Effects of Extracellular Matrices."

21. If a zygote is capable of producing other cells that are truly totipotent, then logically, development cannot proceed until a cell is produced that does not (or cannot) generate a zygote and is therefore not totipotent in the same manner as the original zygote. Stated in a different way, if the first act of a true zygote formed by sperm-egg fusion is to produce other zygotes that are identical to itself, then all zygotes will only and forever produce other zygotes.

22. Bischoff, Parfitt, and Zernicka-Goetz, "Formation of the Embryonic-Abembryonic"; Fujimori et al., "Analysis of Cell Lineage"; Gardner, "Specification of Embryonic Axes"; Gardner and Davies, "Basis and Significance"; Gardner

and Davies, "Investigation of the Origin"; Piotrowska and Zernicka-Goetz, "Role for Sperm"; Plusa et al., "Site of the Previous"; Plusa et al., "First Cleavage."

23. Piotrowska et al., "Blastomeres Arising"; Piotrowska-Nitsche et al., "Four-Cell Stage"; Tabansky et al., "Developmental Bias"; Torres-Padilla et al., "Histone Arginine Methylation."

24. Bell and Watson, "SNAI1 and SNAI2"; Hansis, Grifo, and Krey, "Candidate Lineage Marker"; Held et al., "Transcriptome Fingerprint"; Plachta et al., "Oct4 Kinetics"; Roberts et al., "Transcript Profiling"; Sun et al., "Differential Expression."

25. Galán et al., "Functional Genomics"; Hartshorn et al., "Single-Cell Duplex"; Jędrusik et al., "Role of Cdx2"; May et al., "Multiplex rt-PCR"; Niwa et al., "Interaction between Oct3/4"; Skamagki et al., "Asymmetric Localization"; Wang et al., "Feasibility of Human."

26. Casser et al., "Retrospective Analysis."

27. Schramm and Paprocki, "In Vitro Development"; Shinozawa et al., "Development of Rat"; Tarkowski, Ozdzenski, and Czołowska, "Identical Triplets"; Tarkowski et al., "Individual Blastomeres."

28. Paepe et al., "Human Trophectoderm."

29. Suwińska et al., "Blastomeres of the Mouse."

30. Rossant, "Investigation of the Determinative State."

31. Carr, "Experimental Study"; Tutton and Carr, "Fate of Trophoblast."

32. Condic, "Totipotency," 799.

33. The term "oocyte" strictly refers to an immature female gamete that has not yet completed meiosis. In contrast, an ovum is a fully mature female gamete. These terms tend to be used interchangeably in the literature. Since the fully mature "ovum," which has completed both rounds of meiosis, only exists after sperm-egg fusion, when the resulting cell is more properly considered a zygote or one-cell embryo, I preferentially use the term "oocyte" to refer to female gametes at all stages of maturation.

34. Grøndahl et al., "Dormant and the Fully Competent."

35. Xue et al., "Genetic Programs"; Yan et al., "Single-Cell RNA-Seq."

36. Condic, "Role of Maternal-Effect Genes."

Chapter 3

1. See, for example, Aach et al., "Addressing the Ethical Issues"; Warmflash et al., "Method."

2. Many of the entities listed in table 3.1 are discussed in much greater detail in Condic and Condic, *Human Embryos, Human Beings*, 195–222.

3. The definition of an embryo has been discussed in greater detail in Condic, "Biological Definition."

4. Walker, "Altered Nuclear Transfer"; Schindler, "Response."

5. A "cell" is defined as "a small usually microscopic mass of protoplasm bounded externally by a semipermeable membrane, usually including one or more nuclei and various nonliving products, capable alone or interacting with other cells of performing all the fundamental functions of life, and forming the smallest structural unit of living matter capable of functioning independently." In turn, "life" is defined as "a state of living characterized by capacity for metabolism, growth, reaction to stimuli, and reproduction" (*Merriam-Webster's Medical Dictionary*). Importantly, many complex metabolic processes can be conducted in a test tube using only organic compounds or nonliving cell lysates. Thus, while some metabolic processes persist in a cytoplast for a period of time, cytoplasts are incapable of "performing all of the fundamental functions of life," including growth, reaction to stimuli, and reproduction, and therefore are not living cells.

6. Touati and Wassmann, "How Oocytes Try."

7. Alikani, Schimmel, and Willadsen, "Cytoplasmic Fragmentation."

8. A helpful discussion of the general approach known as systems biology can be found on the National Institutes of Health website: Wanjek, "Systems Biology as Defined by NIH."

9. Condic, "Totipotency."

10. Condic, "Biological Definition," 220–21, with minor modifications.

11. Genomic imprinting is defined as a "genetic alteration of a gene or its expression that is inferred to take place from the observation that certain genes are expressed differently depending on whether they are inherited from the paternal or maternal parent" (*Merriam-Webster's Medical Dictionary*). Imprinted genes are chemically modified in a sperm-specific or egg-specific manner to give different expression patterns.

12. Condic, "Role of Maternal-Effect Genes."

13. For a more comprehensive discussion see Condic, "Embryos and Integration"; Condic and Flannery, "Contemporary Aristotelian Embryology."

14. For a detailed discussion of parthenotes, see Condic and Condic, *Human Embryos, Human Beings*, 195–222.

15. Joza et al., "Essential Role."

16. Berge et al., "Wnt Signaling."

17. Condic, "Totipotency," 802. Internal citations omitted; see references in original.

18. Sozen et al., "Self-Assembly"; Schramm and Paprocki, "In Vitro Development"; Shinozawa et al., "Development of Rat"; Tarkowski, Ozdzenski, and Czołowska, "Identical Triplets." Tarkowski et al., "Individual Blastomeres."

19. Condic, "Role of Maternal-Effect Genes."

20. Recent work has shown that combining stem cells originally derived from embryos with properties similar to TE and ICM can result in aggregates that exhibit many, but not all, of the properties of normal embryos. Sozen et al., "Self-Assembly"; Harrison et al., "Assembly of Embryonic."

21. In *Metaphysics* (1.980a.21), Aristotle held that human beings are ordered by a rational principle, in addition to the nutritive principle shared with plants and the sensory principle shared with animals.

22. Potency will be discussed in greater detail in chapter 5.

23. There are over one hundred billion neurons in the human brain and at least a trillion glial cells. Ongoing classification of neuronal cell types based on function and/or gene expression (see the SciCrunch website, https://scicrunch .org/scicrunch/interlex/view/ilx_0107497) has identified over three hundred distinct classes of neurons (see the "Children" tab). If one considers cells with distinct *operations* (e.g., neurons in the somatosensory cortex that encode information from your right hand, as opposed to your left hand), there are likely to be hundreds of thousands of distinct neuronal-cell types.

24. Condic and Condic, *Human Embryos, Human Beings*, 177–94.

25. Development is the defining feature of embryos at all stages, but in practice, distinguishing zygotes/totipotent cells from other human cell types is more challenging than distinguishing later-stage embryos from collections of cells that are not embryos. For twins arising by blastocyst splitting after the formation of TE and ICM, development is evidenced by the progression of the entity through the appropriate next steps of development—e.g., gastrulation, neurulation, and formation of organs. The status of an entity formed by blastocyst splitting that *does not* undergo the subsequent stages of development would be determined by the same logic presented here for the zygote: an entity generated at a stage of normal development after the formation of TE and ICM can be identified as an embryo by virtue of its ability to produce a full or partial developmental sequence. Thus, when an entity produced by blastocyst splitting arrests and dies, this suggests (but does not prove) it is a nonembryo, while one that advances through (minimally) the next step of development (segregation of parietal endoderm from the ICM), regardless of how far it advances beyond this point, has demonstrated the capacity for development, and is therefore an embryo.

26. Condic, "When Does Human Life Begin? The Scientific Evidence."

27. For example, tumors can form tissues and, in some cases, complex structures such as teeth and eyes. See Devoize et al., "Giant Mature Ovarian Cystic Teratoma"; Sergi et al., "Huge Fetal Sacrococcygeal Teratoma."

28. For a more detailed discussion, see Condic, "Biological Definition," especially figure 1.

29. A minor caveat to this conclusion is the case of developmental variation that is produced by random or stochastic events. Many developmental processes are not exclusively determined by genes or molecules, but involve an element of "chance." An example is an "equivalence group." The cells in such a group have identical developmental potential, yet when one cell of the group randomly differentiates along a particular path, it suppresses this ability in the surrounding cells. Due to such natural, random variation, not all individuals of the same genotype will be exactly identical—yet these are subtle differences that do not affect the overall structure or function of the organism.

30. As noted above, an enucleated cytoplast is arguably not a living cell, and therefore cannot be considered an egg.

31. An epistemological problem is one of knowledge or truth; i.e., discerning whether an entity is an embryo requires a means of accurately determining the potency of the entity, and (in most cases) this can only be determined by observation over time. If it is difficult (or indeed, impossible) to discern potency at the instant an entity is formed, this reflects a limitation in our ability to discern what is true (i.e., our knowledge). In contrast, ontology is concerned with the nature of being. If there were evidence that a human embryo could convert to a nonhuman embryo, this would raise the ontological question of how an entity as a whole could transition between two kinds of beings without any perceptible change in its composition or behavior.

32. The specific nature of any living entity is determined by its substantial form, the organizing principle that logically and metaphysically proceeds and causes the specific molecular composition of the entity. Substantial form will be discussed in more detail in chapter 8 under "An Aristotelian View of the Embryo."

33. Due to the inherent differences in molecular composition between an embryo and a CHM, this is never the case. For example, all of the embryonic events associated with the maternally derived nucleus do not occur in a CHM.

34. Agar, "Embryonic Potential."

35. Findlay et al., "Human Embryo."

36. Rankin, "Can One Be Two?"; Griniezakis and Symeonides, "Twin Conception"; St. John, "And on the Fourteenth Day"; Grobstein, "Biological Characteristics of the Preembryo."

37. Different authors assign the point of "brain function" quite differently, ranging from ten weeks after sperm-egg fusion up to twenty-eight days after birth. For a range of opinions, see Himma, "What Philosophy of Mind"; Penner and Hull, "Beginning of Individual Human Personhood"; Burgess and Tawia, "When Did You First Begin to Feel It?"

38. Byrne, "Use of Anencephalic Newborns as Organ Donors."

39. Himma, "What Philosophy of Mind."

40. Penner and Hull, "Beginning of Individual Human Personhood."

41. Burgess and Tawia, "When Did You First Begin to Feel It?"

42. Kuhse and Singer, *Should the Baby Live?*

43. For example, a child born with a defect in the HTT gene encoding a protein known as *huntingtin* will be normal at birth and will proceed through normal human maturation until approximately forty years of age before exhibiting neurological deficits characteristic of Huntington's disease, including the loss of rational thought. Yet such an individual (1) clearly is a human with a genetic defect, not a nonhuman, and (2) clearly was an individual with Huntington's disease prior to manifesting symptoms. Similarly, a human embryo that undergoes normal maturation yet fails to produce a human brain capable of rational thought is nonetheless a human organism (i.e., a human being) with a defect in a part. Such an entity is clearly not a tumor (i.e., an aggregate of living human cells that shows no global integration of parts, but only cellular organization). If one were to posit that a human entity undergoing development is not a human being, a case would have to be made that it is a novel, nonhuman organism that is nonetheless of human origin and that possesses an intact human genome (a difficult philosophical case to make).

Chapter 4

1. Rankin, "Can One Be Two?"; Griniezakis and Symeonides, "Twin Conception"; St. John, "And on the Fourteenth Day."

2. For simplicity's sake, we will consider only the case of identical twins, and not the case of higher-order identical siblings (triplets, etc.). Even less is known regarding the formation of higher-order identical siblings, but the general process is likely to be similar and the philosophical concerns are identical.

3. Machin, "Non-identical Monozygotic Twins"; Machin, "Familial Monozygotic Twinning."

4. Mateizel et al., "Do ARTs Affect the Incidence"; Knopman et al., "What Makes Them Split?"; Knopman et al., "Monozygotic Twinning."

5. Moore, *Developing Human*; Schoenwolf, *Larsen's Human Embryology*; Sadler, *Langman's Medical Embryology*.

6. Herranz, "Timing of Monozygotic Twinning"; Boklage, "Traces of Embryogenesis."

7. Katayama, Ellersieck, and Roberts, "Development of Monozygotic."

8. Kyono, "Precise Timing of Embryo Splitting."

9. Ziomek, Johnson, and Handyside, "Developmental Potential of Mouse."

10. Eckardt, McLaughlin, and Willenbring, "Mouse Chimeras."

11. Ruffing et al., "Effects of Chimerism."

12. Pera et al., "Gene Expression Profiles"; Brink et al., "Origins of Human Embryonic Stem Cells"; O'Leary et al., "Tracking the Progression"; Varela et al., "Different Telomere-Length Dynamics"; Tang et al., "Tracing the Derivation."

13. Chen et al., "Quantification of Factors."

14. Otsuki et al., "Symmetrical Division."

15. Seshagiri et al., "Cellular and Molecular Regulation"; Cheon et al., "Role of Actin Filaments"; Niimura et al., "Time-Lapse Videomicrographic Observations"; Perona and Wassarman, "Mouse Blastocysts"; Sireesha et al., "Role of Cathepsins"; Sharma et al., "Implantation Serine Proteinases."

16. Yan et al., "Eight-Shaped Hatching."

17. Langendonckt et al., "Atypical Hatching"; Behr and Milki, "Visualization of Atypical Hatching."

18. Knopman et al., "What Makes Them Split?"

19. Yan et al., "Eight-Shaped Hatching."

20. Mateizel et al., "Do ARTs Affect the Incidence"; Nakasuji et al., "Incidence of Monozygotic Twinning"; Sharara and Abdo, "Incidence of Monozygotic Twins"; Milki et al., "Incidence of Monozygotic Twinning"; Kanter et al., "Trends and Correlates."

21. See discussion above under "How Does Twinning Occur?"

Chapter 5

1. Condic and Condic, *Human Embryos, Human Beings*, 77–105.

2. Donceel and Javed, "Immediate Animation."

3. Ford, *When Did I Begin?*

4. Heaney, "Aquinas and the Presence"; Eberl, "Aquinas's Account"; Pasnau, *Thomas Aquinas*; Pasnau, "Souls and the Beginning of Life"; Haldane and Lee, "Aquinas on Human Ensoulment"; Haldane, "Rational Souls."

5. Donceel and Javed, "Immediate Animation," 98.

6. Ford, *When Did I Begin?*, 120.

7. As noted earlier, for simplicity's sake we are considering only the case of twins, but given the argument that identical twinning is likely to occur at the blastocyst stage, even in the case of higher-order identical siblings (triplets, quadruplets, etc.), it is still likely that there are only two totipotent cells in the embryo that go on to split.

8. Condic, "When Does Human Life Begin? The Scientific Evidence."

9. Substantial form and substantial change are discussed in greater detail below (see chapter 8, "An Aristotelian View of the Embryo"). Briefly, in Aristotle's view, substantial form is the ordering principle that causes something to be what it is. When an entity of a specific type is either produced or destroyed (for example, the formation of an organism from two living cells at sperm-egg fusion or the production of a corpse from a living human at the instant of death), a new substantial form must come into existence, and this kind of change is known as a substantial change. For a more comprehensive discussion of substantial form and the embryo, see Condic and Condic, *Human Embryos, Human Beings*.

10. It can be reasonably objected that stem cells in culture spontaneously differentiate into cells with some of the properties of mature human cells (e.g., beating heart cells or blood), and therefore have the proximate active potency to be part of a human being. However, it can also be argued that cells in the laboratory are not, in and of themselves, parts of anything. Because stem cells are not totipotent and cannot intrinsically initiate a program of development, I would conclude that stem cells have the active proximate potency to exhibit characteristics of mature human cells, but only the remote passive potency to participate in development and thereby become part of an existing human being.

11. See discussion in Condic, "Totipotency."

12. Ford, *When Did I Begin?*, 81.

13. Ford, *When Did I Begin?*, 120.

14. Anscombe, "Were You a Zygote?," 41.

15. Gatica et al., "Micromanipulation of Sheep Morulae"; Mitalipov et al., "Monozygotic Twinning"; Reichelt and Niemann, "Generation of Identical Twin Piglets"; Wang, Trounson, and Dziadek, "Developmental Capacity."

16. Morris, Guo, and Zernicka-Goetz, "Developmental Plasticity."

17. Hardy and Handyside, "Cell Allocation"; Rho, Johnson, and Betteridge, "Cellular Composition"; Tao, Reichelt, and Niemann, "Ratio of Inner Cell Mass."

18. Papaioannou, "Lineage Analysis"; Rossant et al., "Interspecific Hybrids"; Zheng et al., "Production of Mouse."

19. Driesch, "Potency of the First Two Cleavage Cells."

20. As noted earlier, an epistemological problem is one of knowledge or truth; i.e., discerning which twin is which requires a means of accurately distinguishing them. If this is difficult (or indeed, impossible), it reflects a limitation in our ability to discern what is true (i.e., our knowledge). In contrast, ontology is concerned with the nature of being. If there were evidence that twins did not have a "separate, concrete existence," this would raise the ontological problem of how two humans could simultaneously be a single human. The (rare) cases

of conjoined twins that function as a single organism raise such an ontological problem.

21. It may be possible (in theory) to establish a criterion to determine which twin is the original. During development, epigenetic modifications to the genome occur both randomly and in an ordered sequence. If the epigenetic state of the original zygote and that of the resulting twins could be compared (something that is not possible using current technology), several outcomes would be possible. Both twins would inevitably be different from the zygote (due to the ordered epigenetic changes occurring during development), yet the twins could either be identical to each other or different, due to one or both twins accruing random epigenetic modifications. Over time, accruing such random modifications is inevitable. Consequently, it would be theoretically possible to determine which twin is most similar to the original zygote and which most divergent. By this criterion, one could argue that the twin with an epigenetic state more similar to that of the original zygote is ontologically identical to the zygote, and the twin that is less similar is the newly generated individual.

22. Technically, only prime matter undergoes substantial change, but an analysis of why this is the case is beyond the scope of the current discussion.

23. Wallace, "Aquinas's Legacy," 188.

24. The case of rape illustrates the fact that "choosing" is not a necessary component of intention. Although the woman does not choose to participate in sexual intercourse, the action is nonetheless intrinsically ordered to reproduction, and therefore (if a pregnancy results) she is rightly considered the mother of the child.

25. The centrality of intention to the reproductive act has a number of implications. Contrast an atypical process of hatching that results in splitting an embryo with a couple having sex purely for their own entertainment. While neither the hatching embryo nor the couple "want" or "mean" to produce a baby, the activities are different. In the case of atypical hatching, the production of a child is a random or unintentional effect of an action that is naturally ordered toward development. In the case of the couple, the action they are performing is intrinsically (naturally) ordered to reproduction; while they may not consciously want a child or "intend" to produce one, their actions nevertheless intend that end. Both of these situations can be contrasted to the case of an experimenter who acts to produce a twin by intentionally spitting an embryo. In this situation, the scientist can legitimately be considered the parent of the twin by virtue of the fact that he has consciously ordered his actions to that end. Consequently, one twin can be seen as the child of the parents donating the embryo for experimentation, and one as the child of the experimenter. This is because the experimenter acts

intentionally to produce a child, and the actions that elicit twinning are ordered to this outcome by design, just as a carpenter who produces a chair from a piece of wood is rightly considered its producer. For a more comprehensive discussion, see Condic and Condic, *Human Embryos, Human Beings*, 77–105.

26. Machin, "Non-identical Monozygotic Twins"; Machin, "Familial Monozygotic Twinning."

27. As discussed above in chapter 2, "How Long Does Totipotency Persist?," if a zygote generates two cells that are truly identical to the original zygote in every respect, development would never proceed, since the daughter cells B and C would also produce zygotes at their first division.

28. Human sperm (excluding the tail) measure roughly five microns by three microns with a volume of approximately 150 cubic microns, making them one of the smallest human cell types. In contrast, a human egg is roughly 100–150 microns in diameter, having a volume of over 1.5 million cubic microns (more than ten-thousand-fold larger than a sperm). Consequently, the oocyte contributes the vast majority of cellular material to the zygote.

29. The degree of genome identity can be calculated in a variety of ways, with most estimates suggesting a more than 95 percent overlap between humans and chimpanzees. See Mikkelsen et al., "Initial Sequence."

30. Ford, *When Did I Begin?*, 122.

31. Kripke, "Identity and Necessity."

32. Brogaard, "Moral Status."

33. Brogaard, "Moral Status," 46.

34. Brogaard, "Moral Status," 47.

35. Brogaard, "Moral Status," 48.

36. Distinguishing the twins is difficult, albeit not necessarily impossible. See discussion above in section "No Corpse," especially n. 21.

37. Brogaard, "Moral Status," 46.

Chapter 6

1. See chapter 4, under "Is the Standard View of Twinning Correct?"

2. For simplicity, we will discuss the case of chimeras occurring in twin gestations although chimeras formed during gestation of multiple siblings are also possible.

3. See, for example, Dutta et al., "Ovotesticular Disorder."

4. For discussion of possible mechanisms of chimera formation in twins see McNamara et al., "Review of the Mechanisms."

5. Lee et al., "Fetal Stem Cell Microchimerism."

6. Chimerism can be even more extensive, with single individuals containing cells from their mother, older siblings, grandmother (via their mother), and any aunts and uncles that preceded their mother in birth order (via their grandmother). Due to the intrinsic limits in the life span of cells, it is unlikely that chimerism extends back more than two generations.

7. Peters et al., "Unusual Twinning Resulting in Chimerism."

8. Scenarios in which the identity of the donor transfers to the host following transplant have been a common theme for horror movies. For example, in *Hands of a Stranger* (1962), hands of a murderer transplanted to a concert pianist transform the pianist into a killer. Such scenarios are fictional (and horrifying) precisely because they defy common sense regarding the continued ontological identity of the host following organ transplant.

9. Condic and Condic, "Defining Organisms."

10. Stevens, "Maternal Microchimerism."

11. These two cases differ in the nature of the disease-causing cells; in sickle cell, the cells themselves are defective, whereas in the case of autoimmune disease, the maternal cells are healthy but cause an immune problem in the context of the fetus.

12. Peters et al., "Unusual Twinning Resulting in Chimerism."

13. See the discussion in chapter 3, under "The Critical Role of Gametes in Establishing Strict Totipotency."

14. Yu et al., "Disputed Maternity."

15. Nistal et al., "Ovotesticular DSD."

16. Yu et al., "Disputed Maternity."

17. See discussion in chapter 4, "Is the Standard View of Twinning Correct?"

18. Otsuki et al., "Symmetrical Division."

19. Ziomek, Johnson, and Handyside, "Developmental Potential of Mouse."

20. Reanimation would require that the substantial form of the embryo somehow persists in the dissociated cells, or in a disembodied state, waiting for the cells to be aggregated. Alternatively, one could argue from a strictly reductionist perspective that both the original and the reaggregated embryos are the same individual, because an individual is nothing more than the sum total of its constituent parts. However, this view is unable to explain the fact that an individual human can remain the same person despite an ever-changing molecular and cellular composition. For further discussion, see Condic and Condic, *Human Embryos, Human Beings*, especially chapter 1.

21. Ruffling et al., "Effects of Chimerism."

22. This is entirely unlike the case of embryo splitting discussed above (see under "No Corpse" in chapter 5), where the demi-embryo that survives splitting continues without pause along the trajectory established by the developmental program of the original embryo.

23. The precise nature of the intersex disorder exhibited by such individuals will vary considerably, depending on the contribution of male and female cells to various body tissues. For example, if the gonads contain exclusively (or even predominantly) male cells, the individual is likely to develop as a phenotypic male, even if the rest of the body consists primarily of cells with a female genotype.

24. During development, the molecular composition of the embryo and the behavior of its cells are constantly changing. Consequently, for multicellular embryos (unlike individual cells), one cannot determine when a new embryo arises based on these criteria alone. Yet the fact that following fusion, the new chimeric, developmental program deflects the trajectory away from the path being followed by either of the original two embryos, and the fact that the molecular basis and the genetic basis of the developmental program are transformed, both strongly indicate a new individual has been formed.

Chapter 7

1. See the beginning of the introduction.

2. These errors are discussed in detail in Condic, "Totipotency," 798.

3. Gilbert, "When Does Personhood Begin?"

4. Condic, "When Does Human Life Begin? The Scientific Evidence."

5. This topic has also been discussed in Condic, "Virtues Beyond," 99–113.

6. United States Library of Congress, *Trials of War Criminals*.

7. World Medical Association, "Declaration of Helsinki."

8. United States Government Publishing Office, 45 C.F.R. 46.

9. European Treaty Series, "Convention for the Protection."

10. World Medical Association, "Declaration of Helsinki," 2191.

11. Veatch, "Abandon the Dead Donor Rule."

12. Caplan and Patrizio, "Beginning of the End."

13. Alternatively, a nonconsequentialist scientist could agree with historical norms and yet still support embryo-destructive research simply because they believe embryos are nonpersons.

14. For discussion, see Pellegrino, "Metamorphosis of Medical Ethics."

15. For example, a recent study from the United Kingdom estimates the rate of return on investment for publicly funded cardiovascular and cancer research at 9 percent and 10 percent, respectively; see Buxton et al., "Medical Research."

16. Tollefsen, *Biomedical Research*, 11.

17. Ugalmugale, *Assisted Reproductive Technology*.

18. A search of the national database for clinical trials (www.clinicaltrials .gov) on February 28, 2017, using the term "stem cell" returned over five thousand nine hundred studies, only thirty-five of which involve stem cells originally derived from human embryos.

19. Grand View Research, "Stem Cells Market Analysis."

20. Guttmacher Institute, "Induced Abortion in the United States."

21. Paulson, "Why the Department of Health."

22. See the compilation of quotations in the appendix.

23. Appendix, scientific literature citation number 7. The authors specify "most" readers since some readers are likely to be identical twins, one of whom arose at some time after sperm-egg fusion.

24. Levin, "Moral Challenge."

25. Condic and Condic, "Appropriate Limits."

26. Catechism of the Catholic Church, 1808.

27. University of Utah, *2016 Annual Financial Report*.

28. National Institutes of Health, "Research Portfolio Online Reporting Tools."

29. It is common practice in the United States to link academic promotion, space allocation, and even faculty salary to the level of grant funding.

30. Catechism of the Catholic Church, 1807.

31. Klavans and Boyack, "Research Focus."

32. An excellent example is a classic and beautiful paper published in 1983 describing the complete embryonic lineage of every cell in a microscopic soil worm; the paper has been cited over two thousand five hundred times in the literature and recognized with the 2002 Nobel Prize in Physiology or Medicine. See Sulston et al., "Embryonic Cell Lineage."

33. Condic and Condic, "Appropriate Limits."

34. Catechism of the Catholic Church, 1805–29.

35. Catechism of the Catholic Church, 1809.

36. Naik, "Mice Are Created from Two Males."

37. Catechism of the Catholic Church, 1806.

38. The following suggestions are adapted from Condic, "Virtues Beyond."

39. Sharp and Foster, "Community Involvement."

40. See, for example, the work of groups such as Community Voices in Medical Ethics and the Southern Methodist University Maguire Center for Ethics and Public Responsibility.

41. Rothwell, *Hidden STEM Economy*.

42. National Science Foundation, "Cultivating Cultures."

Chapter 8

1. Condic and Condic, "Defining Organisms."

2. Condic, "When Does Human Life Begin? The Scientific Evidence."

3. Condic, "Determination of Death."

4. Condic, "Embryos and Integration," 295.

5. For a more comprehensive discussion of the centrality of substantial form, see Condic and Condic, *Human Embryos, Human Beings*, 20–45.

6. An Aristotelian view of the embryo in light of modern scientific evidence is discussed in detail in the following two works: Condic and Condic, *Human Embryos, Human Beings*; Condic and Flannery, "Contemporary Aristotelian Embryology."

7. Condic and Condic, *Human Embryos, Human Beings*, 27.

8. Strictly speaking, substantial form is the source of the rules and principles governing the function of the body; i.e., the "rules and principles" do not exist independently and somehow come together to "create" a substantial form, but rather the substantial form causes these rules and principles to be present in this particular matter.

9. Condic and Condic, *Human Embryos, Human Beings*, 26.

10. In the view of Aristotle and Aquinas, human substantial form is unique. Aquinas held that the human soul is immortal and will persist after the death of the body, a view that is supported by the theological insights of Christianity and other world religions.

11. That is, "forms or fashions," as opposed to "informs," which carries a less active connotation of providing a foundation or an essential principle without being directly responsible for the observed characteristics; e.g., actions may be informed by specific values, but a potter enforms clay to produce a bowl. In contrast to a potter, substantial form organizes (or enforms) *all* aspects of the matter comprising a human, not merely its physical dimensions.

12. President's Council on Bioethics, "Human Cloning and Human Dignity."

13. Strong, "Preembryo Personhood."

14. Cohen, "Family Law."

15. For example, the opinion of Judge Lawrence in *Miller v. American Infertility Group*. See Vick, "Judge Declares Pre-embryo Is Human."

16. *Jeter v. Mayo.*

17. *Jeter v. Mayo*, p. 41.

18. The PubMed database is administered by the National Institutes of Health and is available at the website of the National Center for Biotechnology Information, http://www.ncbi.nlm.nih.gov/pubmed?db=pubmed. A search was

conducted on November 15, 2017, on file with the author, using the terms "preembryo" OR "pre-embryo" OR "pre embryo" OR "preembryos" OR "pre-embryos" OR "pre embryos" OR "preembryonic" OR "pre-embryonic" OR "pre embryonic."

19. A search of the PubMed database was conducted on November 15, 2017, on file with the author, using the terms "zygote" OR "zygotes" OR "zygotic" OR "preimplantation embryo" OR "preimplantation embryos" OR "blastocyst" OR "blastocysts" OR "morula" OR "morulae" OR "morulas."

20. Search conducted March 17, 2017, on file with the author, revealing 1288 citations, with 213 between 2012 and 2016.

21. Grobstein, "Biological Characteristics of the Preembryo."

22. Grobstein, "Biological Characteristics of the Preembryo," 346.

23. Approximately 10 percent of clinical pregnancies miscarry (see Committee on Practice Bulletins—Gynecology, "Early Pregnancy Loss."). While it is often claimed that 70 percent of embryos do not advance to a clinical pregnancy, the actual number is likely to be much lower. Measures of fecundity (the percentage of cases where a couple of known fertility has intercourse during the fertile period and subsequently conceives a child) is between 40 and 60 percent (for review, see Practice Committee of American Society for Reproductive Medicine, "Optimizing Natural Fertility"). Current estimates suggest nonovulation occurs in 26 to 37 percent of cycles (for example see Prior et al., "Ovulation Prevalence"), and data from assisted reproduction clinics using fresh oocytes from donors of known fertility suggest that 15 to 25 percent of oocytes do not fertilize (for example see Solé et al., "How Does Vitrification Affect Oocyte Viability?"). Therefore, anovulation and nonfertilization alone are likely to account for the observed fecundity rates, with the great majority of zygotes proceeding to a clinical pregnancy.

24. Gurdon, Elsdale, and Fischberg, "Sexually Mature Individuals."

25. Grobstein, "Biological Characteristics of the Preembryo," 347.

26. Lu, "Aristotle on Abortion."

27. Kraut, *Aristotle, "Politics" Books VII and VIII.*

28. Lu, "Abortion and Virtue Ethics," 122n23.

29. Lu, "Aristotle on Abortion."

30. Flannery, "Applying Aristotle in Contemporary Embryology"; Condic and Flannery, "Contemporary Aristotelian Embryology."

31. Lu, "Aristotle on Abortion," 57.

32. Lu, "Aristotle on Abortion," 54.

33. Lu, "Aristotle on Abortion," 55.

34. Lu, "Aristotle on Abortion," 61.

35. Lu, "Aristotle on Abortion," 60.

36. Flannery, "Applying Aristotle in Contemporary Embryology."

37. Condic and Flannery, "Contemporary Aristotelian Embryology."

38. The fact that Aristotle considered human life to have unique value appears clear from his view of ethics. For example, he considers murder to be among a handful of actions that are inherently immoral and absolutely prohibited, stating in the *Nicomachean Ethics* (2.6.1107a8–17): "But not every action nor every passion admits of a mean; for some have names that already imply badness, e.g., spite, shamelessness, envy, and in the case of actions adultery, theft, murder; for all of these and suchlike things imply by their names that they are themselves bad, and not the excesses or deficiencies of them. It is not possible, then, ever to be right with regard to them; one must always be wrong. Nor does goodness or badness with regard to such things depend on committing adultery with the right woman, at the right time, and in the right way, but simply to do any of them is to go wrong." (Translation provided by K. Flannery.)

39. See below, under "Human Embryos and Human Value."

40. For a more detailed discussion, see Condic, "When Does Human Life Begin? The Scientific Evidence"; Condic, "Preimplantation Stages."

41. United Nations, Department of Economic and Social Affairs, Convention on the Rights of Persons with Disabilities.

42. Arguments regarding the intrinsic value of embryos are discussed in detail in George and Tollefsen, *Embryo*.

43. This would certainly include scientific experimentation, but it would also include the use of embryos in ART procedures for the noble end of achieving a pregnancy.

BIBLIOGRAPHY

Aach, John, Jeantine Lunshof, Eswar Iyer, and George M. Church. "Addressing the Ethical Issues Raised by Synthetic Human Entities with Embryo-Like Features." *eLife* 6 (March 2017): e20674.

Abad, María Sánchez-Carpintero, Lluc Mosteiro, Cristina Pantoja, Marta Cañamero, Teresa Rayón, Inmaculada Ors, Osvaldo Graña, et al. "Reprogramming in Vivo Produces Teratomas and iPS Cells with Totipotency Features." *Nature* 502, no. 7471 (October 2013): 340–45.

Agar, Nicholas. "Embryonic Potential and Stem Cells." *Bioethics* 21, no. 4 (January 2007): 198–207.

Aguilar, Pablo S., Mary Baylies, André Fleißner, Laura Helming, Naokazu Inoue, Benjamin Podbilewicz, Hongmei Wang, and M. Wong. "Genetic Basis of Cell-Cell Fusion Mechanisms." *Trends in Genetics* 29, no. 7 (July 2013): 427–37.

Alikani, Mina, Tim Schimmel, and Steen M. Willadsen. "Cytoplasmic Fragmentation in Activated Eggs Occurs in the Cytokinetic Phase of the Cell Cycle, in Lieu of Normal Cytokinesis, and in Response to Cytoskeletal Disorder." *Molecular Human Reproduction* 11, no. 5 (May 2005): 335–44.

Anscombe, G. E. M. "Were You a Zygote?" In *Human Life, Action and Ethics: Essays by G. E. M. Anscombe*, edited by Mary Geach and Luke Gormally, 39–44. Charlottesville: Imprint Academic, 2005.

Antczak, Michael, and Jonathan van Blerkom. "Oocyte Influences on Early Development: The Regulatory Proteins Leptin and STAT3 Are Polarized in Mouse and Human Oocytes and Differentially Distributed within the Cells of the Preimplantation Stage Embryo." *Molecular Human Reproduction* 3, no. 12 (December 1997): 1067–86.

Aquinas, Thomas. *Summa Theologica*. Translated by Fathers of the English Dominican Province. Notre Dame, IN: Christian Classics, 1948.

Badria, Layla Francis, Zouhair Odeh Amarin, A. S. Jaradat, H. Zahawi, Amer Gharaibeh, and A. Zobi. "Full-Term Viable Abdominal Pregnancy: A Case

Report and Review." *Archives of Gynecology and Obstetrics* 268, no. 4 (October 2003): 340–42.

Baertschi, Bernard, and Alexandre Mauron. "Moral Status Revisited: The Challenge of Reversed Potency." *Bioethics* 24, no. 2 (February 2010): 96–103.

Behr, Barry R., and Amin A. Milki. "Visualization of Atypical Hatching of a Human Blastocyst In Vitro Forming Two Identical Embryos." *Fertility and Sterility* 80, no. 6 (December 2003): 1502–3.

Bell, Christine Elizabeth, and Andrew J. Watson. "SNAI1 and SNAI2 Are Asymmetrically Expressed at the 2-Cell Stage and Become Segregated to the TE in the Mouse Blastocyst." *PLoS One* 4, no. 12 (December 2009): e8530.

Berge, Derk ten, Wouter Koole, Christophe Fuerer, Matt Fish, Elif Eroglu, and Roeland Nusse. "Wnt Signaling Mediates Self-Organization and Axis Formation in Embryoid Bodies." *Cell Stem Cell* 3, no. 5 (November 2008): 508–18.

Bischoff, Marcus, David-Emlyn Parfitt, and Magdalena Zernicka-Goetz. "Formation of the Embryonic-Abembryonic Axis of the Mouse Blastocyst: Relationships between Orientation of Early Cleavage Divisions and Pattern of Symmetric/Asymmetric Divisions." *Development* 135, no. 5 (March 2008): 953–62.

Boklage, Charles E. "Traces of Embryogenesis Are the Same in Monozygotic and Dizygotic Twins: Not Compatible with Double Ovulation." *Human Reproduction* 24, no. 6 (June 2009): 1255–66.

Brink, Thore C., Smita Sudheer, Doreen Janke, Justyna Jagodzinska, Marc Jung, and James Adjaye. "The Origins of Human Embryonic Stem Cells: A Biological Conundrum." *Cells, Tissues, Organs* 188, nos. 1–2 (2008): 9–22.

Brogaard, Berit. "The Moral Status of the Human Embryo: The Twinning Argument." *Free Inquiry* 23, no. 1 (2002–3): 45–48.

Burgess, John A., and S. A. Tawia. "When Did You First Begin to Feel It?—Locating the Beginning of Human Consciousness." *Bioethics* 10, no. 1 (January 1996): 1–26.

Buxton, Martin, Stephen Hanney, Steve Morris, Leonie Sundmacher, Jorge Mestre-Ferrandiz, Martina Garau, Jon Sussex, et al. "Medical Research: What's It Worth? Estimating the Economic Benefits from Medical Research in the UK." London: UK Evaluation Forum, 2008. https://www.rand.org/pubs/external_publications/EP20080010.html.

Byrne, Paul. "Use of Anencephalic Newborns as Organ Donors." *Paediatrics and Child Health* 10, no. 6. (July 2005): 335–37.

Caplan, Arthur L., and Pasquale Patrizio. "The Beginning of the End of the Embryo Wars." *Lancet* 73, no. 9669 (March 2009): 1074–75.

Carr, David H. "An Experimental Study of Trophoblast Growth in the Lung." *Obstetrics and Gynecology* 50, no. 4 (October 1977): 473–78.

Casser, E., Samuel H. Israel, S. Schlatt, Verena Nordhoff, and M. Boiani. "Retrospective Analysis: Reproducibility of Interblastomere Differences of mRNA Expression in 2-Cell Stage Mouse Embryos Is Remarkably Poor Due to Combinatorial Mechanisms of Blastomere Diversification." *Molecular Human Reproduction* 24, no. 7 (July 2018): 388–400.

Casser, E., Samuel H. Israel, Anika Witten, Katharina Schulte, S. Schlatt, Verena Nordhoff, and M. Boiani. "Totipotency Segregates between the Sister Blastomeres of Two-Cell Stage Mouse Embryos." *Scientific Reports* 7, no. 1 (August 2017): 8299.

Catechism of the Catholic Church. http://www.vatican.va/archive/ccc_css /archive/catechism/p3s1c1a7.htm.

Chen, Sara X., Anna B. Osipovich, Alessandro Ustione, Leah A. Potter, Susan B. Hipkens, Rama D. Gangula, Weiping Yuan, David W. Piston, and Mark A. Magnuson. "Quantification of Factors Influencing Fluorescent Protein Expression Using RMCE to Generate an Allelic Series in the ROSA26 Locus in Mice." *Disease Models and Mechanisms* 4, no. 4 (July 2011): 537–47.

Cheon, Y. P., Myung Chan Gye, Chang Hong Kim, Byung Moon Kang, Yoon Sik Chang, Sung Sun Kim, and Moon Kyoo Kim. "Role of Actin Filaments in the Hatching Process of Mouse Blastocyst." *Zygote* 7, no. 2 (May 1999): 123–29.

Cohen, I. Glenn. "Family Law—Contract—Supreme Court of New Jersey Holds That Preembryo Disposition Agreements Are Not Binding When One Party Later Objects.—J.B. v. M.B., No. A-9-00, 2001 WL 909294 (N.J. Aug. 14, 2001)." *Harvard Law Review* 115, no. 2 (December 2011): 701–8.

Committee on Practice Bulletins—Gynecology. "The American College of Obstetricians and Gynecologists Practice Bulletin no. 150: Early Pregnancy Loss." *Obstetrics and Gynecology* 125, no. 5 (May 2015): 1258–67.

Condic, M. L. "Alternative Sources of Pluripotent Stem Cells: Altered Nuclear Transfer." *Cell Proliferation* 41, suppl. 1 (December 2007): 7–19.

———. "The Basics about Stem Cells." *First Things* 119 (January 2002): 30–34.

———. "A Biological Definition of the Human Embryo." In *Persons, Moral Worth, and Embryos: A Critical Analysis of Pro-choice Arguments*, edited by Stephen Napier, 211–35. New York: Springer, 2011.

———. "Determination of Death: A Scientific Perspective on Biological Integration." *Journal of Medicine and Philosophy* 41, no. 3 (June 2016): 257–78.

———. "Embryos and Integration." In *Life and Learning XXVI: Proceedings of the Twenty-Sixth University Faculty for Life Conference*, edited by Joseph W. Koterski, 295–323. Bronx: Fordham University Press, 2017.

————. "Human Embryology: Science Politics versus Science Facts." *Quaestiones Disputatae* 5, no. 2 (2014): 47–60.

————. "Life: Defining the Beginning by the End." *First Things* 133 (May 2003): 50–54.

————. "Preimplantation Stages of Human Development: The Biological and Moral Status of Early Embryos." In *Is This Cell a Human Being? Exploring the Status of Embryos, Stem Cells and Human-Animal Hybrids*, edited by Antoine Suarez and Joachim Huarte, 25–43. New York: Springer, 2011.

————. "The Role of Maternal-Effect Genes in Mammalian Development: Are Mammalian Embryos Really an Exception?" *Stem Cell Reviews and Reports* 12, no. 3 (June 2016): 276–84.

————. "The Science and Politics of Cloning: What the News Was All About." *On Point*, Charlotte Lozier Institute, May 1, 2013. https://lozierinstitute.org/the-science-and-politics-of-cloning/.

————. "Totipotency: What It Is and What It Is Not." *Stem Cells and Development* 23, no. 8 (April 2014): 796–812.

————. "Virtues beyond a Utilitarian Approach in Biomedical Research." In *Proceedings of the XXII PAV General Assembly*, 99–113. Rome: Libreria Editrice Vaticana, 2017.

————. "When Does Human Life Begin? A Scientific Perspective." Westchester Institute White Paper (Westchester Institute for Ethics and the Human Person) 1, no. 1 (October 2008): 1–18.

————. "When Does Human Life Begin? The Scientific Evidence and Terminology Revisited." *University of St. Thomas Journal of Law and Public Policy* 8, no. 1 (2014): 44–81.

Condic M. L., and S. B. Condic. "The Appropriate Limits of Science in the Formation of Public Policy." *Notre Dame Journal of Law, Ethics, and Public Policy* 17, no. 1. (2003): 157–79.

Condic M. L., and S. B. Condic. "Defining Organisms by Organization." *National Catholic Bioethics Quarterly* 5, no. 2 (2005): 331–53.

Condic, M. L., and S. B. Condic. *Human Embryos, Human Beings: A Scientific and Philosophical Approach*. Washington, DC: Catholic University of America Press, 2018.

Condic, M. L., and Kevin L. Flannery. "A Contemporary Aristotelian Embryology." *Nova and Vetera* (English Edition) 12, no. 2 (2014): 495–508.

Condic, M. L., and M. S. Rao. "Alternative Sources of Pluripotent Stem Cells: Ethical and Scientific Issues Revisited." *Stem Cells and Development* 19, no. 8 (August 2010): 1121–29.

Cook, Michael. "UK Scientists to Push for 28-Day Limit on Cultivation of Embryos." *BioEdge*, December 10, 2016. https://www.bioedge

.org/bioethics/uk-scientists-to-push-for-28-day-limit-on-cultivation-of
-embryos/12122.

Dahab, Amal A., Rahma Aburass, Wasima Shawkat, Reem Babgi, Ola Essa, and
Razaz H. Mujallid. "Full-Term Extrauterine Abdominal Pregnancy: A Case
Report." *Journal of Medical Case Reports* 5 (October 2011): 531.

Devoize, Laurent, Denise Collangettes, Guillaume le Bouëdec, Florence Mishel-
lany, Thierry Orliaguet, Radhouane Dallel, and Martine Baudet-Pommel.
"Giant Mature Ovarian Cystic Teratoma Including More Than 300 Teeth."
*Oral Surgery, Oral Medicine, Oral Pathology, Oral Radiology, and Endodon-
tics* 105, no. 3 (March 2008): 76–79.

Donceel, Joseph F., and S. Javed. "Immediate Animation and Delayed Hominization." *Theological Studies* 31, no. 1 (February 1970): 76–105.

Driesch, Hans. "The Potency of the First Two Cleavage Cells in Echinoderm
Development: Experimental Production of Partial and Double Forma-
tions." In *Foundations of Experimental Embryology*, edited by Benjamin H.
Willier and Jane M. Oppenheimer. New York: Hafner Press, 1974.

Dutta, Deep, K. S. Shivaprasad, R. N. Das, Srabani Ghosh, Uttara Chatterjee,
Subhankar Chowdhury, and Ranen Dasgupta. "Ovotesticular Disorder of
Sexual Development Due to 47,XYY/46,XY/45,X Mixed Gonadal Dysgene-
sis in a Phenotypic Male Presenting as Cyclical Haematuria: Clinical Pre-
sentation and Assessment of Long-Term Outcomes." *Andrologia* 46, no. 2
(March 2014): 191–93.

Eberl, Jason T. "Aquinas's Account of Human Embryogenesis and Recent Inter-
pretations." *Journal of Medicine and Philosophy* 30, no. 4 (August 2005):
379–94.

Eckardt, Sigrid, K. J. McLaughlin, and Holger Willenbring. "Mouse Chimeras
as a System to Investigate Development, Cell and Tissue Function, Disease Mechanisms and Organ Regeneration." *Cell Cycle* 10, no. 13 (2011):
2091–99.

Edwards, Robert Geoffrey, and Christoph Hansis. "Initial Differentiation
of Blastomeres in 4-Cell Human Embryos and Its Significance for Early
Embryogenesis and Implantation." *Reproductive Biomedicine Online* 11,
no. 2 (August 2005): 206–18.

"Embryology: Questions Asked by Lord Alton of Liverpool." Parliamentary
Business, January 8, 2013. UK Parliament website. https://publications
.parliament.uk/pa/ld201213/ldhansrd/text/130108w0001.htm.

European Treaty Series. "Convention for the Protection of Human Rights and
Dignity of the Human Being with Regard to the Application of Biology
and Medicine." *Journal of Medicine and Philosophy* 25, no. 2 (April 2000):
259–66.

Feigin, Michael E., and Craig C. Malbon. "OSTM1 Regulates Beta-Catenin/ Lef1 Interaction and Is Required for Wnt/Beta-Catenin Signaling." *Cell Signal* 20 (2008): 949–57.

Ferrer Colomer, Modesto, and Luis Miguel Pastor. "The Preembryo's Short Lifetime: The History of a Word." *Cuadernos de Bioética: Revista Oficial de la Asociación Española de Bioética y Ética Médica* 23, no. 79 (September–December 2012): 677–94.

Findlay, Jock K., Michael L. Gear, Peter J. Illingworth, Stephen M. Junk, Gillian Kay, A. H. Mackerras, Adrianne Pope, Harald S. Rothenfluh, and Leeanda Wilton. "Human Embryo: A Biological Definition." *Human Reproduction* 22, no. 4 (April 2007): 905–11.

Flannery, Kevin L. "Applying Aristotle in Contemporary Embryology." *Thomist* 67, no. 2 (2003): 249–78.

Ford, Norman M. *When Did I Begin? Conception of the Human Individual in History, Philosophy, and Science.* Cambridge: Cambridge University Press, 1988.

Fujimori, Toshihiko, Yoko Kurotaki, Jun-Ichi Miyazaki, and Yo-ichi Nabeshima. "Analysis of Cell Lineage in Two- and Four-Cell Mouse Embryos." *Development* 130, no. 21 (November 2003): 5113–22.

Gacek, Christopher, M. "Conceiving Pregnancy: U.S. Medical Dictionaries and Their Definitions of Conception and Pregnancy." *National Catholic Bioethics Quarterly* 10 (Autumn 2009): 543–57.

Galán, Amparo, David Montaner, María Poó, Diana Valbuena, Verónica Ruiz, Cristóbal Aguilar, Joaquín Dopazo, and Carlos Simón. "Functional Genomics of 5- to 8-Cell Stage Human Embryos by Blastomere Single-Cell cDNA Analysis." *PLoS One* 5, no. 10 (October 2010), e13615.

Gardner, Richard L. "Specification of Embryonic Axes Begins before Cleavage in Normal Mouse Development." *Development* 128, no. 6 (March 2001): 839–47.

Gardner, Richard L., and Timothy J. Davies. "The Basis and Significance of Pre-patterning in Mammals." *Philosophical Transactions of the Royal Society of London, Series B, Biological Sciences* 358, no. 1436 (August 2003): 1331–38; discussion 1338–39.

Gardner, Richard L., and Timothy J. Davies. "An Investigation of the Origin and Significance of Bilateral Symmetry of the Pronuclear Zygote in the Mouse." *Human Reproduction* 21, no. 2 (February 2006): 492–502.

Gatica, R., M. P. Boland, T. F. Crosby, and I. Gordon. "Micromanipulation of Sheep Morulae to Produce Mono-Zygotic Twins." *Theriogenology* 21, no. 4 (April 1984): 555–60.

George, Robert P., and Christopher Tollefsen. *Embryo: A Defense of Human Life.* New York: Doubleday, 2008.

A. "When Does Personhood Begin?" In *News and Events*. Swarth-lege, 2019. https://www.swarthmore.edu/news-events/when-does ood-begin.

Gilbert, Scott F., Anna L. Tyler, and Emily J. Zackin. *Bioethics and the New Embry-ology: Springboards for Debate*. Sunderland, MA: Sinauer Associates, 2005.

Grand View Research. "Stem Cells Market Analysis by Product (Adult Stem Cells, hESC, Induced Pluripotent Stem Cells), by Application (Regenera-tive Medicine, Drug Discovery), by Technology, by Therapy, and Segment Forecasts, 2018–2025." June 2017. http://www.grandviewresearch.com /industry-analysis/stem-cells-market.

Griniezakis, Archimandrite Makarios, and Nathanael Symeonides. "Twin Con-ception (Didimogenesis) and Ensoulment." *Human Reproduction and Genetic Ethics* 15, no. 1 (2009): 33–37.

Grobstein, Clifford. "Biological Characteristics of the Preembryo." *Annals of the New York Academy of Sciences* 541, no. 1 (October 1988): 346–48.

———. "External Human Fertilization." *Scientific American* 240, no. 6 (June 1979): 57–67.

Grobstein, Clifford, Marilyn Flower, and John Mendeloff. "External Human Fertilization: An Evaluation of Policy." *Science* 222, no. 4620 (October 1983): 127–33.

Grøndahl, Marie Louise, Rehannah Borup, Jonas Vikeså, Erik Hagen Ernst, Claus Y. Andersen, and Karin Lykke-Hartmann. "The Dormant and the Fully Competent Oocyte: Comparing the Transcriptome of Human Oocytes from Primordial Follicles and in Metaphase II." *Molecular Human Reproduction* 19, no. 9 (September 2013): 600–617.

Guć-Šćekić, Marija, Jelena Milasin, Milena Stevanović, Ljubomir Stojanov and Maja S. Djordjevic. "Tetraploidy in a 26-Month-Old Girl (Cytogenetic and Molecular Studies)." *Clinical Genetics* 61, no. 1 (January 2002): 62–65.

Gurdon, John B., T. R. Elsdale, and M. Fischberg. "Sexually Mature Individu-als of Xenopus Laevis from the Transplantation of Single Somatic Nuclei." *Nature* 182, no. 4627 (July 1958): 64–65.

Guttmacher Institute. "Induced Abortion in the United States." January 2018. https://www.guttmacher.org/fact-sheet/induced-abortion-united-states.

Haldane, John, and Patrick Lee. "Aquinas on Human Ensoulment, Abortion and the Value of Life." *Philosophy* 78, no. 304 (April 2003): 255–78.

Haldane, John, and Patrick Lee. "Rational Souls and the Beginning of Life (A Reply to Robert Pasnau)." *Philosophy* 78, no. 306 (October 2003): 532–40.

Hallgrímsson, Benedikt, and Brian Keith Hall, eds. *Epigenetics Linking Genotype and Phenotype in Development and Evolution*. Berkeley: University of Cali-fornia Press, 2011.

Han, Xiaoning, Michael Chen, Fushun Wang, Martha S. Windrem, Su Xing Wang, Steven Shanz, Qiwu Xu, et al. "Forebrain Engraftment by Human Glial Progenitor Cells Enhances Synaptic Plasticity and Learning in Adult Mice." *Cell Stem Cell* 12, no. 3 (March 2013): 342–53.

Hansis, Christoph, James A. Grifo, and Lewis C. Krey. "Candidate Lineage Marker Genes in Human Preimplantation Embryos." *Reproductive Biomedicine Online* 8, no. 5 (May 2004): 577–83.

Hardy, Kate, and Alan H. Handyside. "Cell Allocation in Twin Half Mouse Embryos Bisected at the 8-Cell Stage: Implications for Preimplantation Diagnosis." *Molecular Reproduction and Development* 36, no. 1 (September 1993): 16–22.

Harper, Joyce Catherine, and Sioban B. Sengupta. "Preimplantation Genetic Diagnosis: State of the ART 2011." *Human Genetics* 131, no. 2 (February 2011): 175–86.

Harrison, Sarah Ellys, Berna Sozen, Neophytos Christodoulou, Christos Kyprianou, and Magdalena Zernicka-Goetz. "Assembly of Embryonic and Extraembryonic Stem Cells to Mimic Embryogenesis In Vitro." *Science* 356, no. 6334 (April 2017): eaal1810. https://science.sciencemag.org/content /356/6334/eaal1810.

Hartshorn, Cristina, Judith J. Eckert, Odelya Hartung, and Lawrence J. Wangh. "Single-Cell Duplex RT-LATE-PCR Reveals Oct4 and Xist RNA Gradients in 8-cell Embryos." *BMC Biotechnology* 7 (December 2007): 87. https:// bmcbiotechnol.biomedcentral.com/articles/10.1186/1472-6750-7-87.

Heaney, S. J. "Aquinas and the Presence of the Human Rational Soul in the Early Embryo." *Thomist* 56, no. 1 (January 1992): 19–48.

Held, Eva, Dessie Salilew-Wondim, Matthias Linke, Ulrich Zechner, Franca Rings, Dawit Tesfaye, Karl Schellander, and Michael Hoelker. "Transcriptome Fingerprint of Bovine 2-Cell Stage Blastomeres Is Directly Correlated with the Individual Developmental Competence of the Corresponding Sister Blastomere." *Biology of Reproduction* 87, no. 6 (December 2012): 154. https://academic.oup.com/biolreprod/article/87/6/154,%201 -13/2514068.

Hernández, Javier M., and Benjamin Podbilewicz. "The Hallmarks of Cell-Cell Fusion." *Development* 144, no. 24 (December 2017): 4481–95.

Herranz, Gonzalo. "The Timing of Monozygotic Twinning: A Criticism of the Common Model." *Zygote* 23, no. 1 (February 2015): 27–40.

Himma, Kenneth Einar. "What Philosophy of Mind Can Tell Us about the Morality of Abortion: Personhood, Materialism, and the Existence of Self." *International Journal of Applied Philosophy* 17, no. 1 (2003): 89–109.

Honecker, Friedemann, Hans Stoop, Frank Mayer, C. Carsten Bokemeyer, Diego H. Castrillon, Yun-Fai Chris Lau, Leendert H. J. Looijenga, and J. Wolter Oosterhuis. "Germ Cell Lineage Differentiation in Non-seminomatous Germ Cell Tumours." *Journal of Pathology* 208, no. 3 (February 2006): 395–400.

Huang, Jian, Xiong Jing, Songqing Fan, Zhu Fufan, Ding Yiling, Pi Pixiang, and Xia Xiaomeng. "Primary Unruptured Full Term Ovarian Pregnancy with Live Female Infant: Case Report." *Archives of Gynecology and Obstetrics* 283, suppl. 1 (March 2011): 31–33.

Hupalowska, Anna, Agnieszka Jedrusik, Meng Zhu, Mark T. Bedford, David M. Glover, and Magdalena Zernicka-Goetz. "CARM1 and Paraspeckles Regulate Pre-implantation Mouse Embryo Development." *Cell* 175 (7): 1902–16.e13.

Ingebrigtsen, Ragnvald. "Studies upon the Characteristics of Different Culture Media and Their Influence upon the Growth of Tissue outside of the Organism." *Journal of Experimental Medicine* 16, no. 4 (October 2012): 421–31.

Isah, Aliyu Y., Yakubu Ahmed, Emmanuel I. Nwobodo, and Bissallah Ahmed Ekele. "Abdominal Pregnancy with a Full Term Live Fetus: Case Report." *Annals of African Medicine* 7, no. 4 (December 2008): 198–99.

"IVF Remains in Legal Limbo." Editorial. *Nature* 327 (1987): 87.

Jędrusik, Agnieszka, David-Emlyn Parfitt, Guoji Guo, Maria Skamagki, Joanna B. Grabarek, Martin J. Johnson, Paul Robson, and Magdalena Zernicka-Goetz. "Role of Cdx2 and Cell Polarity in Cell Allocation and Specification of Trophectoderm and Inner Cell Mass in the Mouse Embryo." *Genes and Development* 22, no. 19 (October 2008): 2692–706.

Jeter v. Mayo Clinic Arizona. 121 P.3d 1256, 211 Ariz. 386 (Ct. App. 2005). https://scholar.google.com/scholar_case?case=13185529812469825137.

Johnson, Walter H., Naida M. Loskutoff, Yves Plante, and Keith J. Betteridge. "Production of Four Identical Calves by the Separation of Blastomeres from an In Vitro Derived Four-Cell Embryo." *Veterinary Record* 137, no. 1 (July 1995): 15–16.

Joza, Nicholas A., Santos A. Susin, Eric Daugas, William L. Stanford, Sarah K. R. Cho, Carol Y. J. Li, Takehiko Sasaki, et al. "Essential Role of the Mitochondrial Apoptosis-Inducing Factor in Programmed Cell Death." *Nature* 410, no. 6828 (March 2001): 549–54.

Kanter, Jessica R., Sheree Lynn Boulet, Jennifer Fay Kawwass, Denise J. Jamieson, and Dmitry M. Kissin. "Trends and Correlates of Monozygotic Twinning after Single Embryo Transfer." *Obstetrics and Gynecology* 125, no. 1 (January 2015): 111–17.

Katayama, Mika, Mark R. Ellersieck, and R. Michael Roberts. "Development of Monozygotic Twin Mouse Embryos from the Time of Blastomere Separation at the Two-Cell Stage to Blastocyst." *Biology of Reproduction* 82, no. 6 (June 2010): 1237–47.

Kiernan, John A. "Pre-Embryos." *Nature* 321, no. 6068 (May 1986): 376.

Klavans, Richard, and Kevin W. Boyack. "The Research Focus of Nations: Economic vs. Altruistic Motivations." *PLoS One* 12, no. 1 (January 2017): e0169383.

Knopman, Jaime M., Lewis C. Krey, Jennifer L. Lee, Mary Elizabeth Fino, Akiva P. Novetsky, and Nicole L. Noyes. "Monozygotic Twinning: An Eight-Year Experience at a Large IVF Center." *Fertility and Sterility* 94, no. 2 (July 2010): 502–10.

Knopman, Jaime M., Lewis C. Krey, Cheongeun Oh, Jennifer L. Lee, Caroline McCaffrey, and Nicole L. Noyes. "What Makes Them Split? Identifying Risk Factors That Lead to Monozygotic Twins after In Vitro Fertilization." *Fertility and Sterility* 102, no. 1 (July 2014): 82–89.

Kraut, Richard. *Aristotle, "Politics" Books VII and VIII, Translation with Commentary*. Oxford: Oxford University Press, 1997.

Kripke, Saul A. "Identity and Necessity." In *Naming, Necessity, and Natural Kinds*, edited by Stephen P. Schwartz, 66–101. Ithaca, NY: Cornell University Press, 1977.

Kuhse, Helga, and Peter Singer. *Should the Baby Live? The Problem of Handicapped Infants*. Oxford: Oxford University Press, 1985.

Kyono, Koichi. "The Precise Timing of Embryo Splitting for Monozygotic Dichorionic Diamniotic Twins: When Does Embryo Splitting for Monozygotic Dichorionic Diamniotic Twins Occur? Evidence for Splitting at the Morula/Blastocyst Stage from Studies of In Vitro Fertilization." *Twin Research and Human Genetics: The Official Journal of the International Society for Twin Studies* 16, no. 4 (August 2013): 827–32.

Langendonckt, Anne van, Christine Wyns, P. A. Godin, Dominique Toussaint-Demylle, and Jacques Donnez. "Atypical Hatching of a Human Blastocyst Leading to Monozygotic Twinning: A Case Report." *Fertility and Sterility* 74, no. 5 (November 2000): 1047–50.

Lee, Eddy S. M., George Bou-Gharios, Elke Seppanen, Kiarash Khosrotehrani, and Nicholas M. Fisk. "Fetal Stem Cell Microchimerism: Natural-Born Healers or Killers?" *Molecular Human Reproduction* 16, no. 11 (November 2010): 869–78.

Levin, Yuval. "The Moral Challenge of Modern Science." *New Atlantis* 14 (2006): 32–46.

Liang, Puping, Yanwen Xu, Xiya Zhang, Chenhui Ding, Rui Huang, Zhen Zhang, Jie Lv, et al. "CRISPR/Cas9-Mediated Gene Editing in Human Tri-pronuclear Zygotes." *Protein and Cell* 6, no. 5 (May 2015): 363–72.

Lu, Matthew T. "Abortion and Virtue Ethics." In *Persons, Moral Worth, and Embryos: A Critical Analysis of Pro-choice Arguments*, edited by Stephen Napier, 122–23. New York: Springer, 2011.

———. "Aristotle on Abortion and Infanticide." *International Philosophical Quarterly* 53, no. 1 (2013): 47–62.

Machin, Geoffrey A. "Familial Monozygotic Twinning: A Report of Seven Pedi-grees." *American Journal of Medical Genetics, Part C, Seminars in Medical Genetics* 151C, no. 2 (May 2009): 152–54.

———. "Non-identical Monozygotic Twins, Intermediate Twin Types, Zygosity Testing, and the Non-random Nature of Monozygotic Twinning: A Review." *American Journal of Medical Genetics, Part C, Seminars in Medical Genetics* 151C, no. 2 (May 2009): 110–27.

Magill, Gerard, and William Neaves. "Ontological and Ethical Implications of Direct Nuclear Reprogramming." *Kennedy Institute of Ethics Journal* 19, no. 1 (March 2009): 23–32; discussion 33–40.

Marsden, Hana Robson, Itsuro Tomatsu, and Alexander Kros. "Model Systems for Membrane Fusion." *Chemical Society Reviews* 40, no. 3 (March 2011): 1572–85.

Mateizel, Ileana, Samuel Santos-Ribeiro, Elisa Done, Liesbet Van Landuyt, Hilde van de Velde, Herman J. Tournaye, and Greta Verheyen. "Do ARTs Affect the Incidence of Monozygotic Twinning?" *Human Reproduction* 31, no. 11 (November 2016): 2435–41.

May, Andreas Balser, Roland Kirchner, Helena T. Mueller, Petra Hartmann, Nady El Hajj, Achim Tresch, Ulrich Zechner, Wolfgang Mann, and Thomas Haaf. "Multiplex rt-PCR Expression Analysis of Developmentally Important Genes in Individual Mouse Preimplantation Embryos and Blas-tomeres." *Biology of Reproduction* 80, no. 1 (January 2009): 194–202.

McLaren, Anne. "Pre-Embryos?" *Nature* 328, no. 6125 (July 1987): 10.

McNamara, Helen C., Stefan Charles Kane, Jeffrey M. Craig, Roger V. Short, and Mark P. Umstad. "A Review of the Mechanisms and Evidence for Typi-cal and Atypical Twinning." *American Journal of Obstetrics and Gynecology* 214, no. 2 (February 2016): 172–91.

Merriam-Webster's Medical Dictionary. Springfield, MA: Merriam-Webster, 2019. https://www.merriam-webster.com/medical.

Mikkelsen, Soeren, LaDeana W. Hillier, Evan E. Eichler, Michael C. Zody, David B. Jaffe, Shiaw-Pyng Yang, Wolfgang Enard, et al. "Initial Sequence of the Chimpanzee Genome and Comparison with the Human Genome." *Nature* 437, no. 7055 (September 2005): 69–87.

Milki, Amin A., Sunny Hee Jun, Mary D. Hinckley, Barry R. Behr, Linda C. Giudice, and Lynn Marie Westphal. "Incidence of Monozygotic Twinning with Blastocyst Transfer Compared to Cleavage-Stage Transfer." *Fertility and Sterility* 79, no. 3 (March 2003): 503–6.

Miller, Calum, and Alexander R. Pruss. "Human Organisms Begin to Exist at Fertilization." *Bioethics* 31, no. 7 (September 2017): 534–42.

Mills, Eugene. "The Egg and I: Conception, Identity, and Abortion." *Philosophical Review* 117, no. 3 (July 2008): 323–48.

Mitalipov, Shoukhrat, Richard R. Yeoman, Hung-Chih Kuo, and Don P. Wolf. "Monozygotic Twinning in Rhesus Monkeys by Manipulation of In Vitro–Derived Embryos." *Biology of Reproduction* 66, no. 5 (May 2002): 1449–55.

Moore, Keith L., T. V. N. Persaud, and Mark G. Torchia. *Developing Human: Clinically Oriented Embryology.* 10th edition. Philadelphia: Saunders, 2016.

Morgani, Sophie M., Maurice A. Canham, Jennifer Nichols, Alexei A. Sharov, Rosa Portero Migueles, Minoru S. H. Ko, and Joshua M. Brickman. "Totipotent Embryonic Stem Cells Arise in Ground-State Culture Conditions." *Cell Reports* 3, no. 6 (June 2013): 1945–57.

Morris, Samantha A., Yu Guo, and Magdalena Zernicka-Goetz. "Developmental Plasticity Is Bound by Pluripotency and the Fgf and Wnt Signaling Pathways." *Cell Reports* 2, no. 4 (October 2012): 756–65.

Murray, Henry A. "Physiological Ontogeny: A. Chicken Embryos; VIII, Accelerations of Integration and Differentiation during the Embryonic Period." *Journal of General Physiology* 9, no. 5 (May 1926): 603–19.

Naik, Gautam. "Mice Are Created from Two Males." *Wall Street Journal,* December 10, 2010. http://www.wsj.com/articles/SB1000142405274870 4447604576008031376020012.

Nakamura, Yasuhiro, Michiyo Takaira, Etsuko Sato, Katuichi Kawano, Osamu Miyoshi, and Norio Niikawa. "A Tetraploid Liveborn Neonate: Cytogenetic and Autopsy Findings." *Archives of Pathology & Laboratory Medicine* 127, no. 12 (December 2003): 1612–14.

Nakasuji, Takashi, Hidekazu Saito, Ryuichiro Araki, Aritoshi Nakaza, Akira Nakashima, Akira Kuwahara, Osamu Ishihara, et al. "The Incidence of Monozygotic Twinning in Assisted Reproductive Technology: Analysis Based on Results from the 2010 Japanese ART National Registry." *Journal of Assisted Reproduction and Genetics* 31, no. 7 (July 2014): 803–7.

National Institutes of Health. "Research Portfolio Online Reporting Tools." Last updated June 12, 2018. https://report.nih.gov/award/index.cfm?

National Science Foundation. "Cultivating Cultures for Ethical STEM (CCE STEM): Program Solicitation NSF 15-532." Washington, DC, 2015. https://www.nsf.gov/pubs/2018/nsf18532/nsf18532.htm.

Niimura, Sueo, Tadahiro Ogata, Ayano Okimura, Taro Sato, Yasuhiko Uchiyama, Takeshi Seta, Hiroshi Nakagawa, Kuniaki Nakagawa, and Yuichi Tamura. "Time-Lapse Videomicrographic Observations of Blastocyst Hatching in Cattle." *Journal of Reproduction and Development* 56, no. 6 (December 2010): 649–54.

Nistal, Manuel, Ricardo Alfaro Paniagua, Pilar González-Peramato, and Miguel Reyes-Múgica. "Perspectives in Pediatric Pathology, Chapter 7, Ovotesticular DSD (True Hermaphroditism)." *Pediatric and Developmental Pathology: The Official Journal of the Society for Pediatric Pathology and the Paediatric Pathology Society* 18, no. 5 (September–October 2015): 345–52.

Niwa, Hitoshi, Yayoi Toyooka, Daisuke Shimosato, Dan Strumpf, Kadue Takahashi, Rika Yagi, and Janet Rossant. "Interaction between Oct3/4 and Cdx2 Determines Trophectoderm Differentiation." *Cell* 123, no. 5 (December 2005): 917–29.

O'Leary, T. Michael, Björn Heindryckx, Sylvie Lierman, David van Bruggen, Jelle J. Goeman, Mado Vandewoestyne, Dieter Lucien Daniël Deforce, Susana M. M. Lopes, and Petra De Sutter. "Tracking the Progression of the Human Inner Cell Mass during Embryonic Stem Cell Derivation." *Nature Biotechnology* 30, no. 3 (February 2012): 278–82.

Otsuki, Junko, Yasushi Nagai, Alexander Lopata, Kazuyoshi Chiba, Lubna Yasmin, and Tadashi Sankai. "Symmetrical Division of Mouse Oocytes during Meiotic Maturation Can Lead to the Development of Twin Embryos That Amalgamate to Form a Chimeric Hermaphrodite." *Human Reproduction* 27, no. 2 (February 2012): 380–87.

Paepe, Caroline de, Greet Cauffman, A. Verloes, Johan Sterckx, Paul Devroey, Herman J. Tournaye, Inge Liebaers, and Hilde van de Velde. "Human Trophectoderm Cells Are Not Yet Committed." *Human Reproduction* 28, no. 3 (March 2013): 740–49.

Paepe, Caroline de, M. V. Krivega, Greet Cauffman, Mieke Geens, and Hilde van de Velde. "Totipotency and Lineage Segregation in the Human Embryo." *Molecular Human Reproduction* 20, no. 7 (July 2014): 599–618.

Papaioannou, Virginia E. "Lineage Analysis of Inner Cell Mass and Trophectoderm Using Microsurgically Reconstituted Mouse Blastocysts." *Journal of Embryology and Experimental Morphology* 68 (April 1982): 199–209.

Pasnau, Robert. "Souls and the Beginning of Life (A Reply to Haldane and Lee)." *Philosophy* 78, no. 306 (October 2003): 521–31.

———. *Thomas Aquinas on Human Nature*. Cambridge: Cambridge University Press, 2002.

Paulson, Richard. "Why the Department of Health and Human Services Should Stop Saying Life Begins at Conception." *Los Angeles Times*, October 26, 2017. https://www.latimes.com/opinion/op-ed/la-oe-paulson-when-life-begins -20171026-story.html.

Pellegrino, Edmund D. "The Metamorphosis of Medical Ethics: A 30-Year Retrospective." *Journal of the American Medical Association* 269, no. 9 (March 1993): 1158–62.

Penner, P. S., and Richard T. Hull. "The Beginning of Individual Human Personhood." *Journal of Medicine and Philosophy* 33, no. 2 (April 2008): 174–82.

Pera, Renee A. Reijo, Christopher DeJonge, Nancy L. Bossert, Mylene Yao, Jean Yee Hwa Yang, Narges Bani Asadi, Wing Wong, Connie Wong, and Meri T. Firpo. "Gene Expression Profiles of Human Inner Cell Mass Cells and Embryonic Stem Cells." *Differentiation; Research in Biological Diversity* 78, no. 1 (July 2009): 18–23.

Perona, Rosario, and Paul M. Wassarman. "Mouse Blastocysts Hatch In Vitro by Using a Trypsin-Like Proteinase Associated with Cells of Mural Trophectoderm." *Developmental Biology* 114, no. 1 (March 1986): 42–52.

Peters, H. Elizabeth, Tamar E. König, Marieke O. Verhoeven, Roel Schats, Velja Mijatovic, Johannes C. F. Ket, and Cornelis B. Lambalk. "Unusual Twinning Resulting in Chimerism: A Systematic Review on Monochorionic Dizygotic Twins." *Twin Research and Human Genetics: The Official Journal of the International Society for Twin Studies* 20, no. 2 (April 2017): 161–68.

Piotrowska, Karolina, Florence Wianny, Roger A. Pedersen, and Magdalena Zernicka-Goetz. "Blastomeres Arising from the First Cleavage Division Have Distinguishable Fates in Normal Mouse Development." *Development* 128, no. 19 (October 2001): 3739–48.

Piotrowska, Karolina, and Magdalena Zernicka-Goetz. "Role for Sperm in Spatial Patterning of the Early Mouse Embryo." *Nature* 409, no. 6819 (January 2001): 517–21.

Piotrowska-Nitsche, Karolina, Aitana Perea-Gomez, Seiki Haraguchi, and Magdalena Zernicka-Goetz. "Four-Cell Stage Mouse Blastomeres Have Different Developmental Properties." *Development* 132, no. 3 (February 2005): 479–90.

Plachta, Nicolas, Tobias Bollenbach, Shirley Pease, Scott E. Fraser, and Periklis Pantazis. "Oct4 Kinetics Predict Cell Lineage Patterning in the Early Mammalian Embryo." *Nature Cell Biology* 13, no. 2 (February 2011): 117–23.

Plusa, Berenika, Joanna B. Grabarek, Karolina Piotrowska, David M. Glover, and Magdalena Zernicka-Goetz. "Site of the Previous Meiotic Division Defines Cleavage Orientation in the Mouse Embryo." *Nature Cell Biology* 4, no. 10 (October 2002): 811–15.

Plusa, Berenika, Anna-Katerina Hadjantonakis, Dionne J. Gray, Karolina Piotrowska-Nitsche, Agnieszka Jędrusik, Virginia E. Papaioannou, David M. Glover, and Magdalena Zernicka-Goetz. "The First Cleavage of the Mouse Zygote Predicts the Blastocyst Axis." *Nature* 434, no. 7031 (March 2005): 391–95.

Powers, Carol L. "Pre-existing Condition in the Womb?" Community Voices in Medical Ethics, Community Ethics Committee, October 23, 2018. http://www.medicalethicsandme.org/.

Practice Committee of American Society for Reproductive Medicine in collaboration with Society for Reproductive Endocrinology and Infertility. "Optimizing Natural Fertility: A Committee Opinion." *Fertility and Sterility* 100, no. 3 (September 2013): 631–37.

President's Council on Bioethics. "Human Cloning and Human Dignity: An Ethical Inquiry." Washington, DC: United States Government Printing Office, 2002.

Price Foley, Elizabeth. *The Law of Life and Death.* Cambridge, MA: Harvard University Press, 2011.

Prior, Jerilynn C., Marit Naess, Arnulf Langhammer, and Siri Forsmo. "Ovulation Prevalence in Women with Spontaneous Normal-Length Menstrual Cycles—A Population-Based Cohort from HUNT3, Norway." *PLoS One* 10, no. 8 (August 2015): e0134473.

Rankin, Mark. "Can One Be Two? A Synopsis of the Twinning and Personhood Debate." *Monash Bioethics Review* 31, no. 2 (September 2013): 37–59.

Reichelt, Bernd, and Heiner Niemann. "Generation of Identical Twin Piglets Following Bisection of Embryos at the Morula and Blastocyst Stage." *Journal of Reproduction and Fertility* 100, no. 1 (January 1994): 163–72.

Rho, G. J., Walter H. Johnson, and Keith J. Betteridge. "Cellular Composition and Viability of Demi- and Quarter-Embryos Made from Bisected Bovine Morulae and Blastocysts Produced In Vitro." *Theriogenology* 50, no. 6 (October 1998): 885–95.

Risselada, H. Jelger, and Helmut Grubmüller. "How SNARE Molecules Mediate Membrane Fusion: Recent Insights from Molecular Simulations." *Current Opinion in Structural Biology* 22, no. 2 (April 2012): 187–96.

Roberts, R. Michael, Mika Katayama, Scott R. Magnuson, Michael T. Falduto, and Karen E. O. Torres. "Transcript Profiling of Individual Twin

Blastomeres Derived by Splitting Two-Cell Stage Murine Embryos." *Biology of Reproduction* 84, no. 3 (March 2011): 487–94.

Rossant, Janet. "Investigation of the Determinative State of the Mouse Inner Cell Mass. II: The Fate of Isolated Inner Cell Masses Transferred to the Oviduct." *Journal of Embryology and Experimental Morphology* 33, no. 4 (July 1975): 991–1001.

———. "Postimplantation Development of Blastomeres Isolated from 4- and 8-Cell Mouse Eggs." *Journal of Embryology and Experimental Morphology* 36, no. 2 (October 1976): 283–90.

Rossant, Janet, B. Anne Croy, David A. Clark, and Verne M. Chapman. "Interspecific Hybrids and Chimeras in Mice." *Journal of Experimental Zoology* 228, no. 2 (November 1983): 223–33.

Rosslenbroich, Bernd. "Properties of Life: Toward a Coherent Understanding of the Organism." *Acta Biotheoretica* 64, no. 3 (September 2016): 277–307.

Rothwell, Jonathan. *The Hidden STEM Economy*. Washington, DC: Brookings Institute, 2013. http://www.brookings.edu/~/media/research/files/reports/2013/06/10-stem-economy-rothwell/thehiddenstemeconomy610.pdf.

Ruffing, N. A., Gary B. Anderson, Robert H. Bondurant, W. Bruce Currie, and R. L. Pashen. "Effects of Chimerism in Sheep-Goat Concepti That Developed from Blastomere-Aggregation Embryos." *Biology of Reproduction* 48, no. 4 (April 1993): 889–904.

Sadler, T. W. *Langman's Medical Embryology*. 13th ed. Philadelphia: Lippincott Williams and Wilkins, 2016.

Saito, Shotaro, and Heiner Niemann. "Effects of Extracellular Matrices and Growth Factors on the Development of Isolated Porcine Blastomeres." *Biology of Reproduction* 44, no. 5 (May 1991): 927–36.

Satouh, Yuhkoh, Naokazu Inoue, Masahito Ikawa, and Masaru Okabe. "Visualization of the Moment of Mouse Sperm-Egg Fusion and Dynamic Localization of IZUMO1." *Journal of Cell Science* 125, pt. 21 (November 2012): 4985–90.

Schindler, David L. "A Response to the Joint Statement, 'Production of Pluripotent Stem Cells by Oocyte Assisted Reprogramming.'" *Communio* 32, no. 2 (2005): 370–80.

Schoenwolf, Gary C., Steven B. Bleyl, Philip R. Brauer, and Philippa H. Francis-West. *Larsen's Human Embryology*. 5th ed. Philadelphia: Elsevier Saunders, 2014.

Schramm, R. Dee, and Ann Marie Paprocki. "In Vitro Development and Cell Allocation Following Aggregation of Split Embryos with Tetraploid or Developmentally Asynchronous Blastomeres in Rhesus Monkeys." *Cloning and Stem Cells* 6, no. 3 (2004): 302–14.

Sehgal, Alka, Lajya Devi Goyal, Poonam Goel, and B. S. Sunita. "Full Term Ovarian Pregnancy: A Case Report." *Australian and New Zealand Journal of Obstetrics and Gynaecology* 45, no. 2 (April 2005): 165–66.

Sergi, Consolato M., Volker Ehemann, Bernd Beedgen, O. Linderkamp, and Herwart F. Otto. "Huge Fetal Sacrococcygeal Teratoma with a Completely Formed Eye and Intratumoral DNA Ploidy Heterogeneity." *Pediatric and Developmental Pathology: The Official Journal of the Society for Pediatric Pathology and the Paediatric Pathology Society* 2, no. 1 (January–February 1999): 50–57.

Seshagiri, Polani B., Shubhendu Sen Roy, G. Sireesha, and Rajnish P. Rao. "Cellular and Molecular Regulation of Mammalian Blastocyst Hatching." *Journal of Reproductive Immunology* 83, no. 1–2 (December 2009): 79–84.

Sharara, Fady I., and Galal Abdo. "Incidence of Monozygotic Twins in Blastocyst and Cleavage Stage Assisted Reproductive Technology Cycles." *Fertility and Sterility* 93, no. 2 (February 2010): 642–45.

Sharma, Nitasha, Shiying Liu, Lin Tang, Jackie Irwin, Guoliang Meng, and D. E. Rancourt. "Implantation Serine Proteinases Heterodimerize and Are Critical in Hatching and Implantation." *BMC Developmental Biology* 6 (December 2006): 61.

Sharp, Richard R., and Morris W. Foster. "Community Involvement in the Ethical Review of Genetic Research: Lessons from American Indian and Alaska Native Populations." *Environmental Health Perspectives* 110, suppl. 2 (April 2002): 145–48.

Shaw, R.W., G. Ndukwe, D. Imoedemhe, G. Burford, and R. Chan. "Attempts to Stimulate Multiple Follicular Growth for IVF by Administration of Pulsatile LHRH." *Clinical Reproduction and Fertility* 5, no. 3 (June 1987): 141–51.

Shinozawa, Tadahiro, Atsushi Sugawara, Asako Matsumoto, Y-J Han, I. Tomioka, Kentaro Inai, Hiroshi Sasada, Eiji Kobayashi, Hiromichi Matsumoto, and Eimei Sato. "Development of Rat Tetraploid and Chimeric Embryos Aggregated with Diploid Cells." *Zygote* 14, no. 4 (November 2006): 287–97.

Shukla, V. K., Suman Pandey, Laxmi Pandey, Sumit Kumar Roy, and Madhur P. Vaidya. "Primary Hepatic Pregnancy." *Postgraduate Medical Journal* 61, no. 719 (September 1985): 831–32.

Sireesha, G., Robert W. Mason, M. A. Hassanein, Sarah Tonack, Alexander Navarrete Santos, Bernd Fischer, and Polani B. Seshagiri. "Role of Cathepsins in Blastocyst Hatching in the Golden Hamster." *Molecular Human Reproduction* 14, no. 6 (June 2008): 337–46.

Skamagki, Maria, Krzysztof B. Wicher, Agnieszka Jędrusik, Sujoy Ganguly, and Magdalena Zernicka-Goetz. "Asymmetric Localization of Cdx2 mRNA during the First Cell-Fate Decision in Early Mouse Development." *Cell Reports* 3, no. 2 (February 2013): 442–57.

Solé, Miquel, Josep Santaló, Montse Boada, Elisabet Clúa, Ignacio Rodríguez, F. X. Martinez, Buenaventura Coroleu, P. Barri, and Anna Veiga. "How Does Vitrification Affect Oocyte Viability in Oocyte Donation Cycles? A Prospective Study to Compare Outcomes Achieved with Fresh versus Vitrified Sibling Oocytes." *Human Reproduction* 28, no. 8 (August 2013): 2087–92.

Southern Methodist University. "Cary M. Maguire Center for Ethics and Public Responsibility." https://www.smu.edu/provost/ethics.

Sozen, Berna, Gianluca Amadei, Andy Cox, Ran Wang, Ellen Na, Sylwia Czukiewska, Lia Chappell, et al. "Self-Assembly of Embryonic and Two Extra-Embryonic Stem Cell Types into Gastrulating Embryo-Like Structures." *Nature Cell Biology* 20, no. 8 (August 2018): 979–89.

Stevens, A. Michal. "Maternal Microchimerism in Health and Disease." Best Practice and Research. *Clinical Obstetrics and Gynaecology* 31 (February 2016): 121–30.

St. John, Jeremy. "And on the Fourteenth Day . . . Potential and Identity in Embryological Development." *Monash Bioethics Review* 27, no. 3 (July 2008): 12–24.

Strong, C. Percy. "Preembryo Personhood: An Assessment of the President's Council Arguments." *Theoretical Medicine and Bioethics* 27, no. 5 (2006): 433–53.

Sulston, John E., Einhard Schierenberg, John White, and J. Nichol Thomson. "The Embryonic Cell Lineage of the Nematode Caenorhabditis Elegans." *Developmental Biology* 100, no. 1 (November 1983): 64–119.

Sun, Jian H., Yong Zhang, Bao Ying Yin, Ji Xia Li, Gen Sheng Liu, W. K. Xu, and Shuang Tang. "Differential Expression of Axin1, Cdc25c and Cdkn2d mRNA in 2-Cell Stage Mouse Blastomeres." *Zygote* 20, no. 3 (August 2012): 305–10.

Suwińska, Aneta, Renata Czołowska, Wacław Ożdżeński, and Andrzej K. Tarkowski. "Blastomeres of the Mouse Embryo Lose Totipotency after the Fifth Cleavage Division: Expression of Cdx2 and Oct4 and Developmental Potential of Inner and Outer Blastomeres of 16- and 32-Cell Embryos." *Developmental Biology* 322, no. 1 (October 2008): 133–44.

Tabansky, Inna, Alan B. Lenarcic, Ryan W. Draft, K. Loulier, Derin B. Keskin, Jacqueline Rosains, José Rivera-Feliciano, et al. "Developmental Bias in Cleavage-Stage Mouse Blastomeres." *Current Biology* 23, no. 1 (January 2013): 21–31.

Tachibana, Masahito, Paula Amato, Michelle Sparman, Nuria Marti Gutierrez, Rebecca Tippner-Hedges, Hong Ma, Eunju Kang, et al. "Human

Embryonic Stem Cells Derived by Somatic Cell Nuclear Transfer." *Cell* 153, no. 6 (June 2013): 1228–38.

Tachibana, Masahito, Paula Amato, Michelle Sparman, Joy Woodward, Dario Melguizo Sanchis, Hong Ma, Nuria Marti Gutierrez, et al. "Towards Germline Gene Therapy of Inherited Mitochondrial Diseases." *Nature* 493, no. 7434 (January 2013): 627–31.

Takahashi, Kazutoshi, Koji Tanabe, Mari Ohnuki, Megumi Narita, Tomoko Ichisaka, Kiichiro Tomoda, and Shinya Yamanaka. "Induction of Pluripotent Stem Cells from Adult Human Fibroblasts by Defined Factors." *Cell* 131, no. 5 (November 2007): 861–72.

Tang, Fuchou, Catalin Barbacioru, Siqin Bao, Caroline L. G. Lee, Ellen Nordman, Xiaohui Wang, Kai Qin Lao, and M. Azim Surani. "Tracing the Derivation of Embryonic Stem Cells from the Inner Cell Mass by Single-Cell RNA-Seq Analysis." *Cell Stem Cell* 6, no. 5 (May 2010): 468–78.

Tao, Ta-yao, Bernd Reichelt, and Heiner Niemann. "Ratio of Inner Cell Mass and Trophoblastic Cells in Demi- and Intact Pig Embryos." *Journal of Reproduction and Fertility* 104, no. 2 (July 1995): 251–58.

Tarkowski, Andrzej K., Waclaw Ożdżeński, and Renata Czołowska. "How Many Blastomeres of the 4-Cell Embryo Contribute Cells to the Mouse Body?" *International Journal of Developmental Biology* 45, no. 7 (October 2001): 811–16.

Tarkowski, Andrzej K., Waclaw Ożdżeński, and Renata Czołowska. "Identical Triplets and Twins Developed from Isolated Blastomeres of 8- and 16-Cell Mouse Embryos Supported with Tetraploid Blastomeres." *International Journal of Developmental Biology* 49, no. 7 (2005): 825–32.

Tarkowski, Andrzej K., Aneta Suwińska, Renata Czołowska, and Wacław Ożdżeński. "Individual Blastomeres of 16- and 32-Cell Mouse Embryos Are Able to Develop into Foetuses and Mice." *Developmental Biology* 348, no. 2 (December 2010): 190–98.

Tarkowski, Andrzej K., and Joanna Wróblewska. "Development of Blastomeres of Mouse Eggs Isolated at the 4- and 8-Cell Stage." *Journal of Embryology and Experimental Morphology* 18, no. 1 (August 1967): 155–80.

Thomson, James A., Joseph Itskovitz-Eldor, Sander S. Shapiro, Michelle A. Waknitz, Jennifer J. Swiergiel, Vivienne S. Marshall, and Jeffrey M. Jones. "Embryonic Stem Cell Lines Derived from Human Blastocysts." *Science* 282, no. 5391 (November 1998): 1145–47.

Tollefsen, Christopher. *Biomedical Research and Beyond: Expanding the Ethics of Inquiry.* New York: Routledge Annals of Bioethics, 2008.

Torres-Padilla, Maria-Elena, David-Emlyn Parfitt, Tony Kouzarides, and Magdalena Zernicka-Goetz. "Histone Arginine Methylation Regulates

Pluripotency in the Early Mouse Embryo." *Nature* 445, no. 7124 (January 2007): 214–18.

Touati, Sandra A., and Katja Wassmann. "How Oocytes Try to Get It Right: Spindle Checkpoint Control in Meiosis." *Chromosoma* 125, no. 2 (June 2015): 321–35.

Tutton, D. A., and David H. Carr. "The Fate of Trophoblast Retained within the Oviduct in the Mouse." *Gynecologic and Obstetric Investigation* 17, no. 1 (1984): 18–24.

Ugalmugale, Sumant. *Assisted Reproductive Technology Market Size.* Global Market Insights, June 2016. https://www.gminsights.com/industry-analysis/assisted -reproductive-technology-market.

United Nations, Department of Economic and Social Affairs. Convention on the Rights of Persons with Disabilities (CRPD). A/RES/61/106. December 13, 2006. https://www.un.org/development/desa/disabilities/convention -on-the-rights-of-persons-with-disabilities.html.

United States Government Publishing Office. 45 C.F.R. 46. Washington, DC: United States Government Printing Office, 2015. https://www.gpo.gov /fdsys/pkg/CFR-2000-title45-vol1/content-detail.html.

United States Library of Congress. *Trials of War Criminals before the Nuernberg Military Tribunals under Control Council Law No. 10.* Washington, DC: United States Government Printing Office, 1949. https://www.loc.gov/rr /frd/Military_Law/NTs_war-criminals.html.

University of Utah. *2016 Annual Financial Report.* http://fbs.admin.utah.edu /download/finreport/2016fin.pdf.

Varela, Elisa, Ralph P. Schneider, S. Ortega, and Maria A. Blasco. "Different Telomere-Length Dynamics at the Inner Cell Mass versus Established Embryonic Stem (ES) Cells." *Proceedings of the National Academy of Sciences of the United States of America* 108, no. 37 (September 2011): 15207–12.

Veatch, Robert M. "Abandon the Dead Donor Rule or Change the Definition of Death?" *Kennedy Institute of Ethics Journal* 14, no. 3 (September 2004): 261–76.

Velde, Hilde van de, Greet Cauffman, Herman J. Tournaye, Paul Devroey, and Inge Liebaers. "The Four Blastomeres of a 4-Cell Stage Human Embryo Are Able to Develop Individually into Blastocysts with Inner Cell Mass and Trophectoderm." *Human Reproduction* 23, no. 8 (August 2008): 1742–47.

Vick, K. "Judge Declares Pre-embryo Is Human." *Journal of Biolaw and Business* 8, no. 3 (2005): 52–53.

Vivanco, Luis, Blanca Bartolomé, Montserrat San Martín, and Alfredo Martínez. "Bibliometric Analysis of the Use of the Term Preembryo in Scientific

Literature." *Journal of the American Society for Information Science* 62 (2011): 987–91.

Walker, Adrian J. "Altered Nuclear Transfer: A Philosophical Critique." *Communio* 31, no. 4 (2004): 649–84.

Wallace, William A. "Aquinas's Legacy on Individuation, Cogitation, and Hominization." In *Thomas Aquinas and His Legacy: Studies in Philosophy and the History of Philosophy*, edited by David M. Gallagher, 173–93. Washington, DC: Catholic University of America Press, 1994.

Wang, Chia-Woei, Ding-Shyan Yao, S. J. Horng, Hsiao-Chen Chiu, Chun-Kai Chen, Choonki Lee, Hong-Yuan Huang, H. S. Wang, Y. S. Soong, and Chia Chu Pao. "Feasibility of Human Telomerase Reverse Transcriptase mRNA Expression in Individual Blastomeres as an Indicator of Early Embryo Development." *Journal of Assisted Reproduction and Genetics* 21, no. 5 (May 2004): 163–68.

Wang, Jiaqiang, Leyun Wang, Guihai Feng, Yukai Wang, Yufei Li, Xin Li, Chao Liu, et al. "Asymmetric Expression of LincGET Biases Cell Fate in Two-Cell Mouse Embryos." *Cell* 175, no. 7 (December 2018): 1887–1901.

Wang, Zhaohui J., Alan Trounson, and Marie Dziadek. "Developmental Capacity of Mechanically Bisected Mouse Morulae and Blastocysts." *Reproduction, Fertility, and Development* 2, no. 6 (1990): 683–91.

Wanjek, Christopher. "Systems Biology as Defined by NIH: An Intellectual Resource for Integrative Biology." *NIH Catalyst* 19, no. 6 (November–December 2011). https://irp.nih.gov/catalyst/v19i6/systems-biology-as-defined-by-nih.

Warmflash, Aryeh, Benoit Sorre, Fred Etoc, Eric D. Siggia, and Ali H. Brivanlou. "A Method to Recapitulate Early Embryonic Spatial Patterning in Human Embryonic Stem Cells." *Nature Methods* 11, no. 8 (August 2014): 847–54.

Warnock, Mary. *Report of the Committee of Inquiry into Human Fertilisation and Embryology*. London: Her Majesty's Stationery Office, 1984. https://www.hfea.gov.uk/media/2608/warnock-report-of-the-committee-of-inquiry-into-human-fertilisation-and-embryology-1984.pdf.

Wasserman, David T. "What Qualifies as a Live Embryo?" *American Journal of Bioethics* 5, no. 6 (November–December 2005): 23–25.

Watt, Jessie L., Adly A. Templeton, Ioannis E. Messinis, Laura Bell, Patricia Cunningham, R. O. Duncan. "Trisomy 1 in an Eight Cell Human Pre-Embryo." *Journal of Medical Genetics* 24, no. 1 (January 1987): 60–64.

West, John D., Roslyn R. Angell, Samuel S. Thatcher, John R. Gosden, Nicholas D. Hastie, Anna F. Glasier, and David T. Baird. "Sexing the Human

Pre-embryo by DNA-DNA In-Situ Hybridisation." *Lancet* 1, no. 8546 (June 1987): 1345–47.

World Medical Association. "Declaration of Helsinki: Ethical Principles for Medical Research Involving Human Subjects." *Journal of the American Medical Association* 310, no. 20 (November 2013): 2191–94. http://jama.jamanetwork.com/article.aspx?articleid=1760318.

Wu, Guangming, Lei Lei, and Hans R. Schöler. "Totipotency in the Mouse." *Journal of Molecular Medicine* 95, no. 7 (July 2017): 687–94.

Xiao, Guo-hong, Dun-jin Chen, Xiaofang Sun, Ruo-Qing She, and Y. L. Mai. "Abdominal Pregnancy: Full-Term Viable Baby." *European Journal of Obstetrics, Gynecology, and Reproductive Biology* 118, no. 1 (January 2005): 117–18.

Xue, Zhigang, Kevin Huang, Chaochao Cai, Lingbo Cai, Chunyan Jiang, Yun Feng, Zhenshan Liu, et al. "Genetic Programs in Human and Mouse Early Embryos Revealed by Single-Cell RNA Sequencing." *Nature* 500, no. 7464 (August 2013): 593–97.

Yan, Liying, Mingyu Yang, Hongshan Guo, Lu Yang, Jun Wu, Rong Li, Ping Liu, et al. "Single-Cell RNA-Seq Profiling of Human Preimplantation Embryos and Embryonic Stem Cells." *Nature Structural and Molecular Biology* 20, no. 9 (September 2013): 1131–39.

Yan, Zheng, Hongxing Liang, Li Deng, Hui Long, Hong Chen, Weiran Chai, Lun Suo, et al. "Eight-Shaped Hatching Increases the Risk of Inner Cell Mass Splitting in Extended Mouse Embryo Culture." *PLoS One* 10, no. 12 (December 2015): e0145172.

Yu, H. S., and S. T. Chan. "Effects of Cadmium on Lactate Dehydrogenase Activities in Mouse Pre-Embryos at Various Stages." *Teratology* 34, no. 3 (December 1986): 313–19.

Yu, Neng H., Margot S. Kruskall, Juan José Yunis, Joan H. M. Knoll, Lynne Uhl, Sharon M. Alosco, Marina Ohashi, et al. "Disputed Maternity Leading to Identification of Tetragametic Chimerism." *New England Journal of Medicine* 346, no. 20 (May 2002): 1545–52.

Zhang, Jianping, Fen Li, and Qiu Sheng. "Full-Term Abdominal Pregnancy: A Case Report and Review of the Literature." *Gynecologic and Obstetric Investigation* 65, no. 2 (2008): 139–41.

Zhang, Xinxin, Li Tianda, Linlin Zhang, Liyuan Jiang, Tongtong Cui, Xuewei Yuan, Chenxin Wang, et al. "Individual Blastomeres of 4- and 8-Cell Embryos Have Ability to Develop into a Full Organism in Mouse." *Genetics and Genomics* 45, no. 12 (December 2018): 677–80.

Zheng, Yue-liang, Man-Xi Jiang, Ying-Chun Ouyang, Qing-Yuan Sun, and Da-Yuan Chen. "Production of Mouse by Inter-Strain Inner Cell Mass Replacement." *Zygote* 13, no. 1 (February 2005): 73–77.

Ziomek, Carol Ann, Martin J. Johnson, and Alan H. Handyside. "The Developmental Potential of Mouse 16-Cell Blastomeres." *Journal of Experimental Zoology* 221, no. 3 (July 1982): 345–55.

INDEX

MAUREEN L. CONDIC
is associate professor of neurobiology
at the University of Utah.

Milton Keynes UK
Ingram Content Group UK Ltd.
UKHW050047210724
445729UK00003BA/44